T0255681

Birkhäuser

Oberwolfach Seminars

Volume 46

More information about this series at http://www.springer.com/series/4812

Wilderich Tuschmann • David J. Wraith

Moduli Spaces of Riemannian Metrics

Wilderich Tuschmann
Fakultät für Mathematik
Karlsruher Institut für Technologie
Karlsruhe, Baden-Württemberg
Germany

David J. Wraith
Department of Mathematics and Statistics
National University of Ireland
Maynooth, County Kildare
Ireland

ISSN 1661-237X ISSN 2296-5041 (electronic)
Oberwolfach Seminars
ISBN 978-3-0348-0947-4 ISBN 978-3-0348-0948-1 (eBook)
DOI 10.1007/978-3-0348-0948-1

Library of Congress Control Number: 2015954178

Mathematics Subject Classification (2010): 53C20, 53C21, 55Q05, 57R15

Springer Basel Heidelberg New York Dordrecht London

Printed on acid-free paper

Springer Basel AG is part of Springer Science+Business Media (www.birkhauser-science.com)

Meiner Mutter Maria

For Ann

Contents

Introduction

Any smooth manifold can be equipped with a smooth Riemannian metric, that is, a smoothly varying inner product on each tangent space. A Riemannian metric allows us to do geometry on the manifold, and in particular, it endows the manifold with 'shape'. Any smooth manifold will admit a huge family of Riemannian metrics, and these metrics can display diverse geometric properties. Since each Riemannian manifold (that is, a smooth manifold equipped with a choice of Riemannian metric) has a notion of shape, it makes sense to try and describe the curvature of such an object.

There are three main ways in which the curvature is usually measured. First, there is the sectional curvature, which is a natural generalization to higher dimensions of the notion of Gaussian curvature for surfaces. The Ricci curvature is an average of sectional curvatures, and the scalar curvature can be viewed as an average of Ricci curvatures. Thus the scalar curvature is a weaker notion of curvature than the Ricci curvature, and in turn, the Ricci curvature is a weaker notion than the sectional curvature. These higher dimensional measures of curvature are more difficult to interpret than the Gaussian curvature. For example the sectional curvature controls distance in some sense, whereas the Ricci curvature is closely related to volume. The scalar curvature controls volume only locally.

A fundamental problem in Riemannian geometry is to understand which manifolds admit metrics displaying certain types of curvature characteristics. Of particular importance amongst the possible curvature characteristics are the various sign-based conditions, for example negative sectional curvature, positive Ricci curvature and so on. Existence issues for positive scalar curvature metrics are reasonably well understood. The situation for positive Ricci and positive (or non-negative) sectional curvature metrics is somewhat less clear. The theory of manifolds with negative sectional curvature is well-developed, however the existence question is far from resolved.

For the most part this existence question has been a primary focus of research. However, there is an equally intruiging secondary question. If a manifold admits a given type of metric, how are such metrics distributed among all possible Riemannian metrics on this object? For example are they rare or common?

A particularly natural question concerns connectedness: given two metrics of a given type on a particular manifold, is it possible to continuously deform one metric into the other *through metrics of the same type*, assuming the space of metrics has been equipped with a suitable topology. If the answer to this question is always yes, this shows that the set of metrics having the given type forms a connected subset of the space of all metrics. Thus it makes sense to ask about the connectedness of the subset of 'nice' metrics in whatever context one is working. If this set is not connected, how many connected components are there? Of course, one could also seek to uncover more subtle aspects of the topology.

Instead of working with spaces of metrics, we might instead choose to work with *moduli spaces* of metrics. For any smooth manifold M, we can consider the group of all self-diffeomorphisms of M, $\mathrm{Diff}(M)$. Let $\mathcal{R}(M)$ denote the space of all Riemannian metrics on M equipped with a suitable topology. Then $\mathrm{Diff}(M)$ acts on $\mathcal{R}(M)$ by pulling back metrics. The quotient of $\mathcal{R}(M)$ by this action is a moduli space of metrics. One can similarly form moduli spaces of metrics satisfying the various curvature conditions, as these conditions are invariant under the action of Diff(M). Just as for spaces of metrics, one can pose questions about the topology of these restricted curvature moduli spaces.

Our aim in writing this book is to bring together the key ideas from both recent and classical results on the topology of (moduli) spaces of Riemanian metrics. To the best of our knowledge, this is the first volume to attempt to do so. We feel that this is a timely exercise, as the field has been developing rapidly in recent years, and this looks set to continue into the future.

The book is laid out as follows. In §1 we begin with a brief introduction to the possible topologies on spaces of Riemannian metrics. We will also see that the space of Riemannian metrics can be given the structure of an infinite dimensional manifold. An object of central importance in this book is the s-invariant of Kreck and Stolz. We will meet this first in §5, however a substantial amount of background material is required before being able to present it. This background includes topics such as spin geometry, the Dirac operator and index theory, and we will develop this in §2 and §3. At the end of §3 we will consider index theory obstructions to positive scalar curvature metrics. This allows us in §4 to look at some classical results about spaces of positive scalar curvature metrics due to Hitchin and Carr, (as well as some very recent results in a similar vein). In §6 we look at applications of the Kreck-Stolz s-invariant to moduli spaces of metrics with positive scalar curvature, positive Ricci curvature and both non-negative and positive sectional curvature. In §7 we look at the 'observer moduli space', which is an alternative kind of moduli space for Riemannian metrics which offers certain advantages over the traditional notion. We then provide in §8 a survey of other results on (moduli) spaces of metrics featuring, for the most part, some form of positive curvature. In §9 we turn our attention to (moduli) spaces of metrics on compact manifolds with negative sectional curvature. In §10 we look at non-compact manifolds with complete metrics of non-negative sectional curvature. Here, results about the moduli space of such metrics are established by an analysis of their 'souls', which is a technique quite different in nature to those described for compact manifolds elsewhere in the book. Finally, in §11 we take a brief look at the Klingenberg-Sakai Conjecture, and its implications for spaces of positively pinched metrics. There are two appendices: Appendix A on K-theory, and Appendix B on the Atiyah-Patodi-Singer index theorem.

Note that we only consider (moduli) spaces of Riemannian metrics under sign-based curvature conditions: we do not explore other types of moduli space, such as moduli spaces of Einstein metrics.

This book grew out of lecture notes prepared by the authors for a seminar of the same name given at the Mathematisches Forschungsinstitut Oberwolfach in June 2014. The authors would like to thank the Institute for the invitation to deliver the seminar, and all the staff for their excellent hospitality. We would also like to thank the seminar participants for providing such a stimulating environment.

It is our pleasure to thank Martin Herrmann who assisted in many ways with the preparation of this manuscript; Boris Botvinnik and Mark Walsh for their advice concerning various parts of the text; and Janice Love and Tony Waldron for their technical help.

Last but not least the authors would like to thank Springer Verlag for the speedy correction of the first printing of this volume which, due to a mistake in the book production process, contained serious typographical errors and no illustrations.

Karlsruhe and Maynooth, December 2015. Wilderich Tuschmann and David J. Wraith

1. Spaces of metrics

The aim of this introductory chapter is to present some basic aspects of analysis and topology for the space $\mathcal{R}(M)$ of complete Riemannian metrics on a smooth manifold M. We also consider the corresponding moduli space $\mathcal{M}(M)$, which is the quotient of $\mathcal{R}(M)$ by the action of the diffeomorphism group $\mathrm{Diff}(M)$. In doing so, we we will freely quote material from the references ([BeEb], [Bl], [Cl], [Eb], [FG], [GG], [GM], [Ham], [KrMi]) which we also recommend to the reader for any further in-depth information.

If M is compact, the space $\mathcal{R}(M)$ is a so-called Fréchet manifold. We will begin by introducing this notion.

Definition 1.1. *A seminorm on a vector space F is a real-valued function $||\cdot|| : F \to \mathbb{R}$ such that*

(1) $||f|| \geq 0$ for all vectors $f \in F$;
(2) $||f+g|| \geq 0$ for all vectors f and g;
(3) $||\lambda f|| = |\lambda|\,||f||$ for all vectors f and scalars λ.

A family of seminorms $\{||\cdot||_\alpha : \alpha \in \Lambda\}$ of F defines a unique topology on F such that a sequence (or net) $f_j \to f$ if and only if $||f_j - f||_\alpha \to 0$ for all $\alpha \in \Lambda$.

A *locally convex topological vector space* is a vector space with a topology which arises from a family of seminorms: each seminorm gives rise to an ϵ-neighbourhood of the origin for any $\epsilon > 0$, and by translation we obtain neighbourhoods about each point in the space. These neighbourhoods then form a sub-base for the desired topology, that is, finite intersections of the neighbourhoods form a basis for the topology. (For more on this, see [Ru; §1].) This topology is Hausdorff if and only if $f = 0$ when every $||f||_\alpha = 0$, and it is metrizable if and only if it can be defined by a countable family of seminorms, (in which case one can always use sequences instead of nets). A sequence $\{f_j\}$ in F is a Cauchy sequence if $||f_j - f_k||_\alpha \to 0$ as $j, k \to \infty$ for all $\alpha \in \Lambda$, and F is said to be complete if every Cauchy sequence in F converges to an element of F.

Definition 1.2. *A Fréchet space is a complete metrizable locally convex topological vector space.*

Examples.

(1) Every Hilbert and every Banach space is a Fréchet space, with its topology given by a single norm.
(2) Let M be a closed smooth finite-dimensional manifold, E be a vector bundle over M and $C^\infty(M,E)$ be the vector space of smooth sections of E. Choose Riemannian metrics and connections on the bundles TM and E, let $\nabla^i f$ denote the ith covariant derivative of a section f of E, and set

$$||f||_n = \sum_{i=0}^{n} \sup |\, \nabla^i f(x) \,| .$$

Then $C^\infty(M,E)$, given the topology defined by the sequence of norms $\{||\cdot||_n\}$, is a Fréchet space.

W. Tuschmann, D.J. Wraith, *Moduli Spaces of Riemannian Metrics*,
Oberwolfach Seminars 46, DOI 10.1007/978-3-0348-0948-1_1

Vector addition and scalar multiplication in a Fréchet space are continuous maps, but in general Fréchet spaces are more complicated than Banach spaces. For example, the dual space F^* of a Fréchet space F is itself Fréchet if and only if F is a Banach space, and the vector space $L(E,F)$ of linear maps between two Fréchet spaces E and F is a Fréchet space if and only if F is Banach. On the other hand, the open mapping and Hahn-Banach theorem continue to hold for Fréchet spaces.

Let us now discuss calculus in Fréchet spaces, starting with the notion of derivative. Let E and F be Fréchet spaces, let $U \subset E$ be open, and let $f : U \subset E \to F$ be a continuous map. We define the differential of f at the point $x \in U$ in the direction $v \in E$ to be

$$Df(x)v := \lim_{t \to 0} \left(f(x+tv) - f(x) \right) / t \,.$$

We define f to be differentiable at x in the direction v if this limit exists, and say that f is C^1 if the limit exists for all $x \in U$ and $v \in E$, and if the map $Df : U \times E \to F$ is continuous in both its arguments. Note that Df is a map from the product $U \times E$ to F. (We do not consider it as a map $U \to L(E,F)$, because as mentioned above, $L(E,F)$ is not necessarily a Fréchet space.)

To define the second derivative, take the partial derivative of the map Df in the first component, i.e., take the derivative as x varies over U. This is because Df is linear in the second component, and hence this partial derivative just gives Df again. Thus, the second derivative is a map $D^2 f : U \times E \times E \to F$.

We can iterate the definitions above to define C^k and C^∞ mappings between Fréchet spaces. The chain rule holds for the differential as thus defined. One can define integrals over curves in the usual way, and for such integrals the fundamental theorem of calculus also holds.

Definition 1.3. *A Fréchet manifold modeled on a Fréchet space E is a Hausdorff topological space M with an atlas of coordinates $\{(U_\alpha, \phi_\alpha) | \alpha \in \mathcal{A}\}$, where each $U_\alpha \subset M$ is open and each $\phi_\alpha : U_\alpha \to E$ is a homeomorphism onto its image. Furthermore, if $U_\alpha \cap U_\beta \neq \emptyset$, we require that the transition map $\phi_\alpha \circ (\phi_\beta)^{-1} : \phi_\beta(U_\alpha \cap U_\beta) \to \phi_\alpha(U_\alpha \cap U_\beta)$ be a smooth mapping of Fréchet spaces.*

Tangent spaces and tangent bundles, smooth and C^k differentiable mappings, vector bundles, fibre bundles, and so on are all defined in the category of Fréchet manifolds. (This is analogous to the case of Banach manifolds - compare [KrMi].) Fréchet Lie groups are Fréchet manifolds that are also groups for which the operations of multiplication and taking the inverse are smooth. A good example of a Fréchet Lie group is the diffeomorphism group of a compact manifold.

Let us now consider topologies and manifold structures on spaces of smooth mappings.

We first focus on the space of sections of a smooth fiber bundle F with m-dimensional fibers over a smooth n-dimensional manifold M. One is typically interested in restricting to sections with a certain regularity, say C^k for $0 \leq k \leq \infty$. One can construct manifolds of sections by using the notion of a jet bundle, which is essentially a bundle that contains information about the Taylor expansions of sections of F. The precise definition is as follows.

Suppose we are given two k-times differentiable local sections of F, ϕ and ψ. Suppose further that ϕ and ψ are both defined on an open neighborhood of $p \in M$. We say that ϕ and ψ are k-equivalent at p if $\phi(p) = \psi(p)$ and the following condition holds. Let (x_i, u_α) be coordinates on F around p such that the (x_i) are coordinates on the base manifold M and (u_α) are coordinates in the fiber directions, i.e. the (x_i, u_α) are the coordinates of a local neighbourhood trivialization. We require that for all multi-indices I taking values in $\{1, ..., n\}$ with $1 \leq |I| \leq k$, and for all $1 \leq \alpha \leq m$:

$$\frac{\partial^{|I|}\phi_\alpha}{\partial x^i}(p) = \frac{\partial^{|I|}\psi_\alpha}{\partial x^i}(p).$$

Thus two local sections are k-equivalent at p if and only if their values at p are equal, and also their first k derivatives at p in some local coordinate system around p. The equivalence class containing the local section ϕ is denoted by $(j^k\phi)(p)$, and is called the k-jet of ϕ at p. The equivalence class of a local section ϕ thus consists of all local sections having Taylor expansion up to order k in local coordinates at p equal to that of ϕ.

The set of all k-jets of local sections of F, denoted $J^k(M, F) := \{(j^k\phi)(p) \,|\, p \in M\}$, where ϕ is a local section of F around p, is called the *k-jet bundle.*

The local coordinate system (x_i, u_α) on F gives rise to a local coordinate system on $J^k(M, F)$ in the following way. For a given k-jet over the $\{x_i\}$-coordinate neighbourhood, consider a representative local section ϕ of F. At any point $p \in M$ take the x_i coordinates of p and the u_α coordinates of ϕ, together with the Taylor coefficients up to order k (with respect to the $\{x_i\}$) of all the ϕ_α. The resulting charts can be used to introduce a topology into $J^k(M, F)$. Moreover it is not difficult to see that $J^k(M, F)$ has a natural structure of a smooth, finite-dimensional fiber bundle over both M and F. If F is a vector bundle over M, so is $J^k(M, F)$.

More generally, one can also define k-jets of smooth mappings $\phi \in C^\infty(M, N)$ between arbitrary smooth finite-dimensional manifolds M^m and N^n. To do this, simply view such a map $\phi : M \to N$ as a section of the trivial bundle $M \times N$ over M, and one thus arrives at the smooth manifold of k-jet bundles $J^k(M, N)$. The dimension of $J^k(M, N)$ is given by dim $J^k(M, N) = m + n\frac{(m+k)!}{m!\,k!}$, where $m = \text{dim}M$ and $n = \text{dim}N$.

We will now define the standard topologies on $C^\infty(M, N)$: these are the C^k or smooth compact-open topology, and the C^k or smooth Whitney topology. The C^k/smooth compact open topology is weaker than the corresponding Whitney topology.

We begin by noting that a map $f \in C^\infty(M, N)$ induces a map $j^k f : M \to J^k(M, N)$, which takes each point in $p \in M$ to the k-jet of f at p, $j^k f(p)$. We therefore have a map

$$j^k : C^\infty(M, N) \to C^0(M, J^k(M, N)).$$

Definition 1.4. *The C^k compact-open topology on $C^\infty(M, N)$ is the topology obtained by pulling back the standard (continuous) compact-open topology on $C^0(M, J^k(M, N))$ under j^k. The smooth compact-open topology is the union of the C^k compact-open topologies for all $k \in \mathbb{N} \cup \{0\}$.*

(To say the same thing another way, the C^k compact-open topology is the smallest topology on $C^\infty(M, N)$ with respect to which j^k is continuous, and the smooth compact-open topology is the smallest for which j^k is continuous for all k.)

Definition 1.5. *For each open subset $U \subset J^k(M,N)$, let $S^k(U) := \{\phi \in C^\infty(M,N) : (j^k\phi)(p) \subset U \text{ for all } p \in M\}$. (Note that $S^k(U) = \emptyset$ if the image of U under the jet bundle projection to M is not the whole of M.) The* Whitney C^k *topology on $C^\infty(M,N)$ is then defined as the topology whose basis is given by the sets $S^k(U)$. The* Whitney C^∞ *or* smooth *topology on $C^\infty(M,N)$ is defined to be the topology whose basis is given by $W := \bigcup_{k=0}^\infty W^k$, where W^k denotes the collection of all open subsets of $C^\infty(M,N)$ with respect to the Whitney C^k topology.*

For a norm-based approach to the above definitions, see Hirsch [H; chapter 2] or Rudin [Ru; page 33].

With a little thought, it is not difficult to see that the above topologies can be given alternative descriptions as follows. A sequence of mappings in $C^\infty(M,N)$ converges in the C^k compact-open topology if and only if it converges in the C^k-norm (i.e. C^k-unformly) on every compact subset of M. Similarly, such a sequence converges in the smooth compact-open topology if and only if it converges C^k-uniformly on compact subsets for all k. For this reason, these compact-open topologies are often referred to as *topologies of uniform convergence on compact subsets.* In contrast, if the sequence converges in the C^k Whitney topology, then we have *global* convergence in the C^k norm (i.e. global C^k-uniform convergence - not just on compact subsets), with the obvious modification in the smooth case. We are therefore justified in calling these *topologies of uniform convergence.* We will often use these terms in the sequel, and unless we explicitly state otherwise, when speaking of $C^\infty(M,N)$ as a topological space we will always mean the topology of smooth convergence on compact subsets.

Note that if M is compact, then the Whitney C^k topology and the compact-open C^k topology coincide on the space $C^\infty(M,N)$ for all $0 \leq k \leq \infty$ (compare [KrMi]).

The next result can be found (with full details) in [GG; page 76] and in [KrMi; §42].

Theorem 1.6. *If M is compact, then $C^\infty(M,N)$ is a Fréchet manifold.*

Idea of the proof. Let M be a compact finite-dimensional manifold and let $\pi : F \to M$ be a smooth fibre bundle with finite-dimensional fibres. Then we make a more general claim than in the theorem, namely that the space of smooth sections $C^\infty(M,F)$ is a Fréchet manifold.

Associated to each section ϕ we have a vector bundle over M called the vertical tangent bundle to F at ϕ, which we denote by $F^v(\phi)$. Its fibre at a point $p \in M$ consists of all the tangent vectors to F at $\phi(x)$ which lie in the null space of the derivative of the projection map π. It is then easy to construct a diffeomorphism from a neighborhood of the zero section of the vector bundle $F^v(\phi)$ to a neighborhood of the image of ϕ in F which takes fibres into fibres over the same point. This provides a one-to-one correspondence between sections near zero in the Fréchet space $C^\infty(M, F^v(\phi))$ and sections near ϕ in $C^\infty(M,F)$, and these maps serve as our coordinate charts. The coordinate transition functions are then vector bundle maps. □

Things tend, however, to become much more delicate when M is not compact. But with an appropriate modification of topologies and a more general notion of manifold, Kriegl and Michor have worked out a satisfying theory which then shows that $C^\infty(M,N)$ is a 'manifold' even for non-compact M and possibly infinite-dimensional N. Since we will not need to consider the Riemannian geometry and manifold structure of $C^\infty(M,N)$ too much in what follows, we refer the interested reader to the comprehensive book [KrMi] and the further references given there.

We now come to the central object of this book, the space of all Riemannian metrics (or all Riemannian geometries) on a given smooth manifold, and its quotient by the diffeomorphism group.

Let M be an n-dimensional smooth manifold, let S^2T^*M denote the second symmetric power of the cotangent bundle of M, and let $C^\infty(M,S^2T^*M)$ be the real vector space of smooth symmetric $(0,2)$ tensor fields on M.

By the previous arguments, $C^\infty(M,S^2T^*M)$ is a Fréchet space if M is compact, and in general we can always topologize $C^\infty(M,S^2T^*M)$ and its subsets with, for example, the C^k or smooth topology of uniform convergence on compact subsets.

Definition 1.7. *The space $\mathcal{R}(M)$ of all complete Riemannian metrics on M is the subspace of $C^\infty(M,S^2T^*M)$ consisting of all sections which are complete Riemannian metrics on M, equipped with the smooth topology of uniform convergence on compact subsets.*

Notice first (compare [FM]) that $\mathcal{R}(M)$ is a convex cone in $C^\infty(M,S^2T^*M)$, i.e., if $a,b>0$ and $g_1,g_2\in\mathcal{R}(M)$, then $ag_1+bg_2\in\mathcal{R}(M)$. Thus $\mathcal{R}(M)$ is contractible, and in particular path-connected.

This latter fact is a principal reason for working on $\mathcal{R}(M)$ with the smooth topology of uniform convergence on compact subsets, for if M is noncompact, then $\mathcal{R}(M)$ equipped with the smooth topology (i.e. the smooth Whitney topology) will in general have an infinite number of path components. Indeed metrics with different geometry at infinity will not lie in the same path component of $\mathcal{R}(M)$.

Let us now consider the moduli space of complete Riemannian metrics. As before, let M denote a smooth finite-dimensional manifold and $\mathrm{Diff}(M)$ be the group of self-diffeomorphisms of M (which, for compact M, is a Fréchet Lie group). Then $\mathrm{Diff}(M)$ acts on $\mathcal{R}(M)$ by pulling back metrics, i.e., one has the action

$$\mathrm{Diff}(M)\times\mathcal{R}(M)\to\mathcal{R}(M),\quad (g,\phi)\mapsto\phi^*(g)\,.$$

Definition 1.8. *The moduli space $\mathcal{M}(M)$ of complete Riemannian metrics on M is the quotient of $\mathcal{R}(M)$ by the above action of the diffeomorphism group $\mathrm{Diff}(M)$, equipped with the quotient topology.*

Notice that usually $\mathrm{Diff}(M)$ will not act freely on $\mathcal{R}(M)$. For instance, if there is an effective action of a finite or compact subgroup $G\subset\mathrm{Diff}(M)$ on M, one can average each Riemannian metric on M over G, and the resulting metric will be fixed by G. Due to the fact that different Riemannian metrics may have isometry groups of different dimension, the moduli space $\mathcal{M}(M)$ will in general not have any kind of manifold structure.

To avoid this situation, a different moduli space is sometimes considered. (See also §7.)

Definition 1.9. *Let M be a connected smooth manifold, $x_0\in M$ a point of M, and $\mathrm{Diff}_{x_0}(M)$ the subgroup of $\mathrm{Diff}(M)$ which consists of all self-diffeomorphisms ϕ of M which fix x_0 and whose differential at x_0 equals the identity map of $T_{x_0}M$. Then $\mathcal{M}_{x_0}(M):=$*

$\mathcal{R}(M)/\mathrm{Diff}_{x_0}(M)$ *is called the observer moduli space of complete Riemannian metrics on* M.

In the following chapters, we will try to relate and explain (most of) the important results on subspaces of $\mathcal{R}(M)$, $\mathcal{M}(M)$, and $\mathcal{M}_{x_0}(M)$, defined by curvature conditions, that are known at the time of writing.

2. Clifford algebras and spin

The aim of this chapter is to develop the concept of a spin group and certain related ideas such as spin structures, spinor representations and spinor bundles. These all play a crucial role in positive scalar curvature geometry, which we will explore in subsequent sections. The importance of these concepts for our purposes is due to the fact that 'spin geometry' provides the setting in which we can define and analyze a certain first order linear differential operator called the Dirac operator. It is this operator which turns out to be intimately related to the scalar curvature. As the name suggests, the Dirac operator arose from work of the physicist Paul Dirac. Leaving the physics on one side, we can motivate the mathematics behind this operator by posing the following question: can we find a first order differential operator D whose square is equal to the Laplacian? In $\mathbb{R}^3$ for example, this amounts to solving the equation

$$D^2 = (\alpha\partial/\partial x + \beta\partial/\partial y + \gamma\partial/\partial z)^2 = -(\partial^2/\partial x^2 + \partial^2/\partial y^2 + \partial^2/\partial z^2).$$

Thus we are required to find coefficient functions α, β and γ satisfying the relations

$$\alpha^2 = \beta^2 = \gamma^2 = -1;$$

$$\alpha\beta = -\beta\alpha; \quad \alpha\gamma = -\gamma\alpha; \quad \beta\gamma = -\gamma\beta.$$

A moment's thought shows that these relations cannot be satisfied by scalar functions. On the other hand they *can* be solved if we allow the coefficients to be *matrices*, and working within certain matrix algebras turns out to be the right setting in which to understand this problem. These matrix algebras are called 'Clifford algebras', and we will begin below by defining these. (Note that it is not obvious from the definition that Clifford algebras are matrix algebras.) The concepts of spin groups and spin geometry will then emerge naturally from Clifford algebras.

The material in this chapter (and the next) is presented in detail in the book [LM]. We therefore present only an outline of the key ideas and results, referring the interested reader to sections I and II of [LM] for the technical details.

§2.1 *Clifford algebras*

The real Clifford algebra Cl_n is formed as follows. Consider the tensor algebra

$$T_n := \mathbb{R} \oplus \mathbb{R}^n \oplus (\mathbb{R}^n \otimes \mathbb{R}^n) \oplus (\mathbb{R}^n \otimes \mathbb{R}^n \otimes \mathbb{R}^n) \oplus (\mathbb{R}^n \otimes \mathbb{R}^n \otimes \mathbb{R}^n \otimes \mathbb{R}^n) \oplus \ldots.$$

Thus elements of T_n are sums of tensor products of elements in $\mathbb{R}^n$.

Within this algebra is an ideal I_n generated by elements of the form $v \otimes v + |v|^2.1$ for $v \in \mathbb{R}^n$. The Clifford algebra Cl_n is given by $Cl_n := T_n/I_n$. This means that the elements of Cl_n are essentially sums of products of elements in $\mathbb{R}^n$ subject to the relation that $v^2 = -|v|^2$ for all $v \in \mathbb{R}^n$, where we have suppressed the tensor product symbol.

W. Tuschmann, D.J. Wraith, *Moduli Spaces of Riemannian Metrics*,
Oberwolfach Seminars 46, DOI 10.1007/978-3-0348-0948-1_2

Applying this relation to a vector $v + w$ we see that $vw + wv = |v|^2 + |w|^2 - |v + w|^2$. In particular for orthonormal vectors α, β we have $\alpha^2 = \beta^2 = -1$, $\alpha\beta = -\beta\alpha$. Notice that these are precisely the sort of relations which arise in the 'Dirac problem' above, which justifies our claim that Clifford algebras are the right setting in which to view that problem.

Clearly, like the tensor algebra, the Clifford algebra is a *graded* algebra, graded by lengths of products of vectors. Moreover there is an obvious analogy with the exterior algebra $\Lambda^*\mathbb{R}^n$, which is defined as T_n/E_n where E_n is the ideal generated by elements of the form $v \otimes v$, i.e. so that $v^2 = 0$ for all $v \in \mathbb{R}^n$. Clearly the subspace $\Lambda^p\mathbb{R}^n$ corresponds to the products of length p in Cl_n, and we have an isomorphism $Cl_n \cong \Lambda^*\mathbb{R}^n$ of graded vector spaces - but not of algebras.

Note that we can define the complex Clifford algebra $\mathbb{C}l_n$ in the same way, just substituting complex scalars for real and using the complex inner product $\langle \sum z_i e_i, \sum w_j e_j \rangle = \sum z_i w_i$.

It is not difficult to see that Cl_n (respectively $\mathbb{C}l_n$) has the structure of a real (respectively complex) vector space of dimension 2^n. A basis is given by

$$\{1, e_1, ..., e_n\} \cup \{e_{i_i} \cdot e_{i_2} \cdot ... \cdot e_{i_k} \,|\, i_1 < i_2 < ... < i_k,\ 2 \leq k \leq n\}$$

where $\{e_i\}$ is the standard basis for $\mathbb{R}^n$ (or $\mathbb{C}^n$). This choice of basis determines a linear isomorphism with $\mathbb{R}^{2^n}$ (or $\mathbb{C}^{2^n}$), and using this we can introduce a topology onto the Clifford algebra which makes the isomorphism into a homeomorphism. It is easy to see that all operations in the Clifford algebra are continuous with respect to this topology, and thus Clifford algebras are topological algebras in a natural way.

It turns out that Clifford algebras - up to algebra isomorphism - are actually all familiar matrix algebras. Let $F(n)$ denote the algebra of $(n \times n)$-matrices over F, where F can be $\mathbb{R}$, $\mathbb{C}$ or $\mathbb{H}$. For small values of n we have the following identifications:

	1	2	3	4	5	6	7	8
Cl_n	$\mathbb{C}$	$\mathbb{H}$	$\mathbb{H} \oplus \mathbb{H}$	$\mathbb{H}(2)$	$\mathbb{C}(4)$	$\mathbb{R}(8)$	$\mathbb{R}(8) \oplus \mathbb{R}(8)$	$\mathbb{R}(16)$
$\mathbb{C}l_n$	$\mathbb{C} \oplus \mathbb{C}$	$\mathbb{C}(2)$	$\mathbb{C}(2) \oplus \mathbb{C}(2)$	$\mathbb{C}(4)$	$\mathbb{C}(4) \oplus \mathbb{C}(4)$	$\mathbb{C}(8)$	$\mathbb{C}(8) \oplus \mathbb{C}(8)$	$\mathbb{C}(16)$

These isomorphisms are not obvious but give useful insights into the Clifford algebras, and in particular to *Clifford modules*. Note that the first few of these isomorphisms can be figured out by hand, and the rest inductively using certain periodicity relations: 8-periodic in the real case $Cl_{n+8} \cong Cl_n \otimes_{\mathbb{R}} Cl_2$, and 2-periodic in the complex case $\mathbb{C}l_{n+2} \cong \mathbb{C}l_n \otimes_{\mathbb{C}} \mathbb{C}l_2$. Using some basic isomorphisms of matrix groups in conjunction with the periodicity isomorphisms, it is not difficult to see that in all cases *the complex Clifford algebras are just complexifications of the real,* i.e. $\mathbb{C}l_n \cong Cl_n \otimes \mathbb{C}$. Therefore for many purposes it suffices to consider the real algebras, and just complexify where necessary.

A *Clifford module* is a vector space which admits a (left, say) linear action from some Clifford algebra: $Cl_n \times V \to V$. For an element $\sigma \in Cl_n$ and $v \in V$, forming a 'product' $\sigma \cdot v$ is called *Clifford multiplication.*

Every Clifford module is either irreducible or splits into a direct sum of irreducible submodules. Using the identification of Clifford algebras with matrix algebras above gives an easy way to describe the irreducible Clifford modules.

Theorem 2.1.1. *(1) If a Clifford algebra is isomorphic to a matrix algebra $F(n)$, then (up to equivalence) there is a unique irreducible Clifford module, namely F^n, with the canonical action of $F(n)$ on F^n given by left matrix multiplication. (2) If a Clifford algebra is isomorphic to a matrix algebra $F(n) \oplus F(n)$, then (up to equivalence) there are two irreducible Clifford modules, both F^n, but with the Clifford algebra action given by the canonical action of $F(n)$ on F^n by one of the factors in $F(n) \oplus F(n)$ with the other factor acting trivially.*

Notice in the table above that there are two irreducible Clifford modules in dimensions 3 mod 4 in the real case, and in every odd dimension in the complex case. In all other dimensions there is a unique irreducible Clifford module. *This is a general phenomenon which holds in all dimensions.* It will be of great significance later.

We next consider splittings of the Clifford algebra. There are two ways of doing this: a way which works for all Clifford algebras, and a more important phenomenon which only works in dimensions 0 mod 4 in the real case and in all even dimensions in the complex case.

Every Clifford algebra has a splitting into *even* and *odd* parts: $Cl_n = Cl_n^0 \oplus Cl_n^1$ (and similarly in the complex case). As a vector space Cl_n^0 is spanned by the even length products of vectors, and Cl_n^1 is spanned by the odd length products. It is easy to see that Cl_n^0 is a subalgebra, but Cl_n^1 is not; moreover $Cl_n^i \cdot Cl_n^j \subset Cl^{i+j \text{ mod } 2}$. An alternative way of viewing this splitting is to consider the algebra isomorphism $\alpha : Cl_n \to Cl_n$ induced by the map $-\mathrm{id} : \mathbb{R}^n \to \mathbb{R}^n$. It is easy to see that $\alpha^2 = \mathrm{id}$, and that Cl_n^0 and Cl_n^1 are respectively just the $+1$ and -1 eigenspaces of α.

For any n we have an algebra isomorphism $Cl_{n-1} \simeq Cl_n^0$, which is induced by the linear map $\mathbb{R}^{n-1} \to Cl_n^0$ given by mapping $e_i \mapsto e_n \cdot e_i$ where $\{e_1, ..., e_n\}$ is the standard orthonormal basis for $\mathbb{R}^n$, and $\mathbb{R}^{n-1} = \mathrm{Span}\{e_1, ..., e_{n-1}\}$. In the complex case we have the analogous isomorphism $\mathbb{C}l_{n-1} \cong \mathbb{C}l_n^0$.

The second way of splitting the Clifford algebra involves the use of a volume element $\omega := e_1 \cdot ... \cdot e_n$. Note that this element is independent of the choice of (oriented) basis once an orientation for $\mathbb{R}^n$ has been fixed. For $\mathbb{C}l_n = Cl_n \otimes \mathbb{C}$ we introduce a complex volume element $\omega_C := i^{\lfloor (n+1)/2 \rfloor} e_1 \cdot ... \cdot e_n$. This means that the complex volume form is real in dimensions 0 and 3 mod 4. More specifically we have $\omega_C = -\omega$ in dimensions 3 and 4 mod 8, and $\omega_C = \omega$ in dimensions 7 and 0 mod 8.

These volume elements have the following properties:

$$\omega^2 = 1 \text{ if } n \equiv 0, 3 \text{ mod } 4;$$

$$e \cdot \omega = (-1)^{n-1} \omega \cdot e \text{ for } e \in \mathbb{R}^n;$$

$$\omega_C^2 = 1 \text{ for all } n;$$

$$e \cdot \omega_C = (-1)^{n-1}\omega_C \cdot e \text{ for all } n.$$

When the square of the volume element is 1, this shows that (left) multiplication by the volume elements splits the Clifford algebra in these dimensions into a sum of the ± 1-eigenspaces. We write $Cl_n = Cl_n^+ \oplus Cl_n^-$ in the real case, and similarly in the complex case. This is only a vector space splitting in general.

The fact that ω commutes with elements of $\mathbb{R}^n$ in dimensions 3 mod 4 in the real case shows that multiplication by ω is an algebra isomorphism (or a module isomorphism in the case of Clifford modules) in these dimensions. Thus the splitting into $Cl_n^\pm$ is a splitting into *subalgebras*, each of which are ideals in Cl_n. These subalgebras are isomorphic, as can be seen from the above table.

In dimensions 0 mod 4 notice that multiplication of $Cl_n^\pm$ by $e \in \mathbb{R}^n$ swaps the subspaces, i.e.

$$\mathbb{R}^n \times Cl_n^\pm \to Cl_n^\mp.$$

This phenomenon occurs for the splitting $Cl_n^0 \oplus Cl_n^1$ under multiplication by elements of Cl_n^1.

In the complex case we getting splittings in all dimensions. When n is odd ω_C is central, and the splitting is into isomorphic subalgebras, as illustrated in the table. When n is even we have a vector space splitting $\mathbb{C}l_n^\pm$ only, and this exhibits the same swapping behaviour under Clifford multiplication by elements of $\mathbb{C}^n$ as dimensions 0 mod 4 in the real case, namely

$$\mathbb{C}^n \times \mathbb{C}l_n^\pm \to \mathbb{C}l_n^\mp.$$

Note that if we compare the splittings $Cl_n = Cl_n^0 \oplus Cl_n^1$ and $Cl_n = Cl_n^+ \oplus Cl_n^-$ in the dimensions where the latter splittings occur, we find that the two splittings are 'diagonal' with respect to one another. To see this note that the map $\alpha : Cl_n \to Cl_n$ swaps the $Cl_n^\pm$, since n is odd and therefore $\alpha(\omega) = -\omega$. This means that for $\phi \in Cl_n^+$ we have $\omega\alpha(\phi) = -\alpha(\omega)\alpha(\phi) = -\alpha(\omega\phi) = -\alpha(\phi)$. Thus $\alpha(Cl_n^+) = Cl_n^-$, and vice versa. From this we see that $Cl_n^0 = \{\phi \oplus \alpha(\phi) \,|\, \phi \in Cl_n^+\} \subset Cl_n^+ \oplus Cl_n^-$. Similarly in the complex case.

The splitting of Clifford algebras has implications for Clifford modules. Firstly note that a Cl_n-module W is said to be $\mathbb{Z}_2$-graded if W admits a vector space splitting $W = W^0 \oplus W^1$ such that the Clifford multiplication satisfies $Cl^i \cdot W^j \subset W^{i+j}$, with the indices interpreted modulo 2. Notice that this makes both W^0 and W^1 modules for the even part Cl_n^0 (but not for the full algebra). Not every Clifford module will admit such a splitting: a pre-requisite is that W has a splitting into two Cl_{n-1}-modules (since $Cl_n^0 \cong Cl_{n-1}$). However there is natural equivalence between the categories of $\mathbb{Z}_2$-graded modules for Cl_n and ungraded modules for Cl_{n-1}. In one direction we just take W^0; in the other we form $Cl_n \otimes_{Cl_{n-1}} W$ for an ungraded Cl_{n-1}-module W, with the action of Cl_n being the usual left action on the first factor. This Cl_n-module splits as $(Cl_n^0 \otimes_{Cl_{n-1}} W) \oplus (Cl_n^1 \otimes_{Cl_{n-1}} W)$, giving a $\mathbb{Z}_2$-grading. These $\mathbb{Z}_2$-graded modules have a useful link with K-theory, which is outlined in Appendix A.

Now consider a module V for the real Clifford algebra Cl_n. When the Clifford algebra volume element ω satisfies $\omega^2 = 1$, i.e. when $n \equiv 0, 3 \bmod 4$, we get a splitting of V into the ± 1-eigenspaces for multiplication by ω: $V^{\pm} = (1 \pm \omega)/2 \cdot V$.

In the case $n = 3 \bmod 4$ we have $V = V^+ \oplus V^-$, and any non-zero element $e \in \mathbb{R}^n$ has the effect (via Clifford multiplication) $e : V^{\pm} \to V^{\pm}$, as a consequence of the centrality of ω. Moreover both $V^{\pm}$ are still modules for Cl_n. In particular this means that if V is irreducible, then either $V = V^+$ or $V = V^-$, i.e. V must be just a single eigenspace. Thus there are precisely two inequivalent (consider the action of ω!) irreducible modules for Cl_n when $n \equiv 3 \bmod 4$, as indicated in the theorem above.

When $n = 0 \bmod 4$ we get $e : V^+ \to V^-$ and $e : V^- \to V^+$. (We see this phenomenon when studying Dirac operators.) For this reason the $V^{\pm}$ are not modules for Cl_n, however since ω commutes with Cl_n^0 we see that they are in fact modules for $Cl_n^0 \cong Cl_{n-1}$. In the case that V is an irreducible Cl_n-module, the resulting Cl_{n-1} modules $V^{\pm}$ are precisely the two inequivalent irreducible modules discussed in the $n \equiv 3 \bmod 4$ case above. Notice also that for $n \equiv 0 \bmod 4$ the splitting $V^{\pm}$ gives V a $\mathbb{Z}_2$-graded structure. In fact there are two such structures, depending on whether we label V^+ as V^0 or V^1.

In the case of a module W for a complex Clifford algebra $\mathbb{C}l_n$, analogous observations apply: we obtain a splitting $W = W^+ \oplus W^-$ for all n, and when n is odd this is a splitting into submodules. For n even Clifford multiplication gives a map $\mathbb{C}^n \times W^{\pm} \to W^{\mp}$, where $W^{\pm}$ can be viewed as modules for the subalgebra $\mathbb{C}l_n^0 \simeq \mathbb{C}l_{n-1}$. Similar comments about irreducibilty apply: for n odd and W irreducible, we have two possibilities for W depending on whether the complex volume element acts as $+1$ or -1. For n even we obtain a splitting of an irreducible module W into the two irreducible modules for $\mathbb{C}l_{n-1} \cong \mathbb{C}l_n$.

§2.2 *The Spin groups*

It is well-known that any element of $\mathrm{SO}(n)$ can be expressed as a product of an even number of reflections across hyperplanes. This is not difficult to see: by a change of basis any rotation matrix can be put into block diagonal form where the non-trivial diagonal blocks are all (2×2)-rotation matrices. The claim will then follow if we can show that any two-dimensional rotation can be expressed as a product of two reflections. An elementary calculation in plane geometry shows that a reflection through a line making an angle of θ with the positive x-axis followed by a reflection through a line making an angle ϕ is equivalent to an anticlockwise rotation through an angle $2(\theta - \phi)$. This also demonstrates the extent to which the representation of rotations by reflections - even in two dimensions - is non-unique.

Given a rotation in $\mathbb{R}^n$, express this as a product of reflections $r_1....r_m$. For each r_i we can find a unit vector $v_i \subset \mathbb{R}^n$ orthogonal to the hyperplane of reflection (and therefore determining both the hyperplane and the reflection). Thus we could express the rotation by the string $v_1....v_m$. Notice that we could replace any of the v_i by $-v_i$ without changing the corresponding rotation.

Let us interpret the string $v_1....v_m$ as a product inside the Clifford algebra Cl_n. By the above observation, the element $-v_1...v_m$ also represents the same rotation (any minus signs can be carried to the front since we are now working in an algebra). Thus there are two expressions giving the same rotation. As noted above, the initial choice of reflections $r_1....r_m$ is not unique, however the remarkable thing is that the corresponding Clifford algebra representations will be equal (in the algebra) to either $v_1....v_m$ or $-v_1....v_m$. (This can be seen, for example, from the short exact sequence below.) *Thus the Clifford algebra is an appropriate setting in which to consider rotations.*

Consider the multiplicative subgroup of Cl_n generated by the collection of unit vectors in $\mathbb{R}^n$. (We need to keep things real here as we are investigating $\mathrm{SO}(n)$.) Within this subgroup we can consider the even length products of unit vectors. It is not difficult to see that this has the structure of a multiplicative group.

Definition 2.2.1. *For each $n \in \mathbb{N}$, the spin group Spin(n) is given by*

$$Spin(n) := \{v_1 \cdot ... \cdot v_r \,|\, v_i \in \mathbb{R}^n \text{ is a unit vector and } r \text{ is even}\}.$$

Since for any unit vector v we have $v \cdot v = -1$ and $(-v) \cdot v = +1$, we see that every spin group contains the scalars ± 1. (It is not difficult to see that these are the only pure scalars in $\mathrm{Spin}(n)$: the pure scalars must form a multiplicative subgroup of $\mathrm{Spin}(n)$, and so must be a multiplicative subgroup of the non-zero real numbers. Working with respect to an orthonormal basis for $\mathbb{R}^n$ we see that the scalar term of any Clifford product of unit vectors must lie in $[-1, 1]$, and the only multiplicative group within this interval is $\{\pm 1\}$.) Identifying $\{\pm 1\}$ with $\mathbb{Z}_2$, the correspondence between $\mathrm{Spin}(n)$ and $\mathrm{SO}(n)$ can be expressed via the following short exact sequence of groups

$$0 \to \mathbb{Z}_2 \to \mathrm{Spin}(n) \to \mathrm{SO}(n) \to 0.$$

Recall from §2.1 that the Clifford algebra is a topological space in a natural way. The induced topology then makes $\mathrm{Spin}(n)$ into a topological group, and with respect to this topology the map $\pi : \mathrm{Spin}(n) \to \mathrm{SO}(n)$ in the above exact sequence is a continuous homomorphism. Thus $\mathrm{Spin}(n)$ is a topological double cover of $\mathrm{SO}(n)$, and since $\pi_1(\mathrm{SO}(n)) \cong \mathbb{Z}_2$ for $n \geq 3$, we see that $\mathrm{Spin}(n)$ is *simply-connected* for $n \geq 3$. Moreover, as the universal cover of a manifold, we see that $\mathrm{Spin}(n)$ is also a manifold, and moreover a Lie group with respect to the smooth structure it inherits from $Cl_n \cong \mathbb{R}^{2^n}$ (or equivalently by pulling back the smooth structure from $\mathrm{SO}(n)$ via π).

In low dimensions note that there are certain group isomorphisms which express $\mathrm{Spin}(n)$ in terms of other Lie groups. We have:

$$\mathrm{Spin}(1) = \mathbb{Z}_2;$$

$$\mathrm{Spin}(2) = S^1;$$

$$\mathrm{Spin}(3) \cong \mathrm{SU}(2) \cong \mathrm{Sp}_1 \cong S^3;$$

$$\mathrm{Spin}(4) \cong \mathrm{Spin}(3) \times \mathrm{Spin}(3);$$

$$\mathrm{Spin}(5) \cong \mathrm{Sp}(2);$$

$$\mathrm{Spin}(6) \cong \mathrm{SU}(4).$$

§2.3 *Spin structures*

Consider a fibre bundle E with fibre F and base B. If $\{U_\alpha\}$ is a covering of B by open balls, we can view E as constructed from products $U_\alpha \times F$ by gluing the fibres over overlap regions $U_\alpha \cap U_\beta$ using maps $\theta_{\alpha\beta} : U_\alpha \cap U_\beta \to G$, where $G \subset \mathrm{Diff}F$ is a fixed Lie group (the 'structure' group) and where the maps $\theta_{\alpha\beta}$ must satisfy the relation $\theta_{\alpha\beta}\theta_{\beta\gamma}\theta_{\gamma\alpha} = 1$. Thus

$$E = \Big(\coprod_\alpha U_\alpha \times F\Big)/\sim$$

where $(u_\alpha, f) \sim (u_\beta, f')$ if and only if $u_\alpha = u_\beta$ and $\theta_{\alpha\beta}(u_\alpha)(f) = f'$.

Given bundle gluing data as above, we can form a new bundle from the same data by replacing the fibre F by the group G, and taking the action of the structural group G on the fibre G to be the usual left multiplication. This creates the associated 'principal G-bundle' P_G. Notice that since we can multiply G by itself from both left and right, the bundle P_G admits a global right G-action. (The left action was 'used up' by the construction process.)

Given the principal G-bundle $G \to P_G \to B$ we can recover the original bundle (up to bundle equivalence) via the *associated bundle construction.* This involves quotienting the product $P_G \times F$ by the equivalence relation $\sim$ defined by $(p, f) \sim (pg^{-1}, g \cdot f)$ for any $g \in G$. Thus the original bundle is associated to the principal bundle P_G by the construction $(P_G \times F)/\sim$.

More generally, given a manifold X on which G acts, we can form an X-bundle associated to P_G in the same way, as $(P_G \times X)/\sim$, where $(p, x) \sim (pg^{-1}, g \cdot x)$. In this way we can create many new bundles sharing the 'same' gluing information.

Consider a manifold M^n. To say that M is orientable is equivalent to saying that the tangent bundle has structure group $\mathrm{SO}(n)$, or equivalently that TM has an associated principal $\mathrm{SO}(n)$-bundle $P_{SO(n)}$.

The manifold M is said to have a *spin structure* if it is orientable and the principal $\mathrm{SO}(n)$-bundle associated to the tangent bundle $P_{SO(n)} \to M$ can be lifted to a principal $\mathrm{Spin}(n)$-bundle $P_{Spin(n)} \to M$ via a map $\xi : P_{Spin(n)} \to P_{SO(n)}$, such that ξ restricts to a double covering map on each fibre, and which is equivariant in the sense that $\xi(pg) = \xi(p)\pi(g)$, where $p \in P_{Spin(n)}$, $g \in \mathrm{Spin}(n)$ and $\pi : \mathrm{Spin}(n) \to \mathrm{SO}(n)$ is the standard double covering homomorphism.

We might therefore think of the existence of a spin structure for M in terms of M satisfying an enhanced orientability condition.

We can characterise both the orientability and spin conditions topologically using the Stiefel-Whitney classes $w_1 \in H^1(M;\mathbb{Z}_2)$ and $w_2 \in H^2(M;\mathbb{Z}_2)$. These provide a useful criterion for checking the existence of spin structures. A manifold M is orientable if and only if $w_1 = 0 \in H^1(M;\mathbb{Z}_2)$, and M is spin if and only if both $w_1 = 0$ and $w_2 = 0$. (These facts are far from obvious. They can be established by studying cohomology exact sequences which arise from certain fibrations - see [LM; §II.1].)

Note that spin structures (that is principal Spin-bundles $\xi : P_{Spin(n)} \to P_{SO(n)}$ for given $P_{SO(n)}$) are not in general unique. In fact the number of spin structures turns out to be in one-to-one correspondence with the elements of $H^1(P_{SO(n)};\mathbb{Z}_2)$ for which the restriction to the fibre of $P_{SO(n)}$ is non-zero. Assuming X is connected, the spin structures on a bundle E over X (not necessarily the tangent bundle) are indexed by $H^1(X;\mathbb{Z}_2)$. (See [LM; page 81] for details.)

There are other characterisations of spin structures. For example in dimensions at least 5, a manifold is spin if and only if every compact orientable embedded surface has trivial normal bundle. (It suffices to consider only embedded 2-spheres if the manifold is simply-connected.) For comparison, the manifold is orientable if the restriction of the normal bundle to any embedded circle is trivial (which is equivalent to the restriction of the tangent bundle being trivial since S^1 is parallelisable).

We can extend the notion of spin structure naturally to any orientable vector bundle by lifting (if possible) the corresponding principal $\mathrm{SO}(n)$-bundle to a principal $\mathrm{Spin}(n)$-bundle. Similar characterisations apply in terms of the vanishing of Stiefel-Whitney classes, and also in terms of pull-back bundles over embeddings of orientable surfaces in the base being trivial.

Many of the most obvious examples of manifolds admit spin structures. For example all spheres S^n admit spin structures. This is easy to see for $n \geq 3$ as then $H^1(S^n;\mathbb{Z}_2)$ and $H^2(S^n;\mathbb{Z}_2)$ are both zero. All Lie groups are spin. This can be seen from the fact that the tangent bundle of every Lie group is trivial, which forces the vanishing of all Stiefel-Whitney classes. Among the projective spaces $\mathbb{R}P^n$ is spin if and only if $n \equiv 3$ mod 4, $\mathbb{C}P^n$ is spin if and only if n is odd, and $\mathbb{H}P^n$ is spin for all n. Note that any product of spin manifolds is again spin, but this is not in general true for bundles. For example there is a unique non-trivial S^3-bundle over S^2 which is known to be non-spin, despite both S^2 and S^3 being spin.

§2.4 *Spinor bundles*

Consider the Clifford algebra $Cl_n = T_n/I_n$. If we transform $\mathbb{R}^n$ by an element of $\mathrm{SO}(n)$, this induces a natural transformation of Cl_n since it transforms the tensor algebra T_n whilst preserving the inner product and hence the ideal I_n. Thus we obtain a group homomorphism (representation) $\gamma_n : \mathrm{SO}(n) \to \mathrm{Aut}(Cl_n)$.

Let E be the total space of an orientable n-plane vector bundle. There is an associated principal bundle $P_{SO(n)}(E)$. To this principal bundle we can associate a bundle with fibre Cl_n using the representation γ_n above:

$$P_{SO(n)}(E) \times_{\gamma_n} Cl_n := P_{SO(n)}(E) \times Cl_n / \sim$$

where $(p,\sigma) \sim (pg^{-1}, \gamma_n(g)(\sigma))$ for $g \in \mathrm{SO}(n)$ and $\sigma \in Cl_n$. This is the *Clifford bundle* associated to E. We will denote this bundle of Clifford algebras by $Cl(E)$. (Remember that if we forget its multiplicative structure, Cl_n is just a vector space and therefore $Cl(E)$ is a vector bundle - albeit a vector bundle of a special kind.) Notice that Cl_n has $\mathbb{R}^n$ as a vector subspace, and that the action γ_n restricted to $\mathbb{R}^n$ is just the usual left action of $\mathrm{SO}(n)$ on $\mathbb{R}^n$. Thus we see that E itself is a sub-bundle of $Cl(E)$. In particular this means that we can regard sections of E as sections of $Cl(E)$, i.e. $\Gamma(E) \subset \Gamma(Cl(E))$.

If E is non trivial, we will not be able to canonically identify individual fibres of $Cl(E)$ with Cl_n. However within each fibre we have a well-defined multiplication between elements (as γ_n preserves multiplication in Cl_n). As a consequence we see that $\Gamma(Cl(E))$ *has the structure of an algebra.*

Now suppose that E is the total space of an n-plane vector bundle with spin structure, for example the tangent bundle of a spin manifold. In this case we are able to define very special vector bundles associated to E called *spinor bundles.*

Consider the associated principal spin bundle $P_{Spin(n)}$, and take any Clifford module V. Now $\mathrm{Spin}(n) \subset Cl_n$, and so the action of the Clifford algebra on V, $\rho : Cl_n \times V \to V$, restricts to give an action of $\mathrm{Spin}(n)$ on V. We can then use this action to form new associated bundles $P_{Spin(n)} \times_\rho V$. Such bundles are called *spinor bundles.* It is not difficult to see that the bundle $P_{Spin(n)} \times_\rho V$ is a bundle of modules over the bundle of algebras $Cl(E)$, in the sense that each fibre in the spinor bundle is a module over the corresponding algebra fibre in $Cl(E)$. (This point is explained in detail in [LM; page 97].)

The following fact will be fundamental in the definition of Dirac operators:

Theorem 2.4.1. *Let $S(E)$ be a spinor bundle associated to E (either real or complex). Then the space of sections $\Gamma(S(E))$ is a left module over the space of sections $\Gamma(Cl(E))$.*

Note that the sections of spinor bundles are usually referred to as *spinors.*

Looking at the table in §2.1 we see, for example, that the *real* Clifford algebras in dimensions 1 and 5 have the structure of a complex algebra. By the periodicity phenomenon, this is true more generally in dimensions congruent to 1 and 5 modulo 8. Consequently the irreducible modules for these algebras are naturally complex vector spaces. Since complex multiplication clearly commutes with the Clifford action on these modules, we see that the corresponding irreducible (real!) spinor bundles actually have the structure of complex vector bundles, and that Clifford multiplication is complex linear. Consequently the space of (real!) spinors is naturally a complex vector space (of infinite dimension). We can make similar statements about quaternionic structures for spinor bundles and spaces of spinors in dimensions 2, 3 and 4 modulo 8.

Typically we will be interested in irreducible spinor bundles, that is, spinor bundles for which V is an irreducible Clifford module. As we saw in §2.1, in many dimensions there is a unique irreducible Clifford module. Thus in these cases we have a unique irreducible spinor bundle for each principal spin bundle. In the other cases, there are two such bundles. In our discussion of spinor bundles we have so far been implicitly assuming that V is a real Clifford module, since the spin groups arise naturally in a real Clifford algebra context. However we can also work with complex modules V for the complex Clifford algebra $\mathbb{C}l_n = Cl_n \otimes \mathbb{C}$. Recall that there is a unique irreducible module for the complex Clifford algebra in all even dimensions, and therefore corresponding to this we have a unique irreducible complex spinor bundle for each principal spin bundle in these dimensions. We will denote such bundles by $\mathbb{S}$. In the next chapter we will see that these play an important role in the theory of positive scalar curvature.

As a simple example of a *reducible* spinor bundle, we can consider Cl_n as a left module over itself. Thus we get a spinor bundle $P_{Spin(n)} \times_\rho Cl_n$ where ρ here is the left action of Cl_n on itself restricted to $\mathrm{Spin}(n)$. Note that this is not the same as the bundle $Cl(E)$. In fact

$$Cl(E) = P_{Spin(n)} \times_{Ad} Cl_n,$$

where $Ad : \text{Spin}(n) \to \text{Aut}(Cl_n)$ denotes the adjoint action $Ad(g)(\sigma) = g\sigma g^{-1}$.

Just as for Clifford algebras and Clifford modules, Clifford and spinor bundles are subject to splittings. As $\text{Spin}(n) \subset Cl_n^0$, we see that the canonical action of $\text{Spin}(n)$ on Cl_n respects the splitting $Cl_n = Cl_n^0 \oplus Cl_n^1$, as does the adjoint action. It follows (from the observation about the adjoint action) that we obtain a corresponding bundle splitting $Cl(E) = Cl(E)^0 \oplus Cl(E)^1$. In the case of splittings by the real volume element in dimensions 0 and 3 mod 4, and by the complex volume element in all even dimensions, we obtain corresponding bundle splittings by observing that the volume element in the Clifford algebra gives a global volume section in $Cl(E)$ respectively $\mathbb{C}l(E)$. Consequently in these dimensions we have $Cl(E) = Cl(E)^+ \oplus Cl(E)^-$, $S(E) = S(E)^+ \oplus S(E)^-$, and similarly in the complex case. Sections of S^+ (respectively S^-) are referred to as positive (respectively negative) spinors. In dimensions 0 mod 4 in the real case (and in all even dimensions in the complex case), fibrewise Clifford multiplication yields the following maps:

$$Cl^0(E) \times S^{\pm}(E) \to S^{\pm}(E);$$

$$Cl^1(E) \times S^{\pm}(E) \to S^{\mp}(E).$$

In particular this means that multiplying positive (respectively negative) spinors by sections of $Cl^1(E)$ produces negative (respectively positive) spinors in these dimensions.

3. Dirac operators and index theorems

In this chapter we explore the theory of obstructions to positive scalar curvature metrics. Such obstructions will form the foundation stone for future chapters concerning the topology of (moduli) spaces of positive scalar curvature metrics. These obstructions arise from studying the so-called 'Dirac operator' on spinor bundles (see §3.1). More specifically, we will see that the existence of a positive scalar curvature metric forces the vanishing of a quantity called the 'index' (see §3.2) associated to the Dirac operator. Via the Atiyah-Singer index theorem (§3.3), the vanishing of the index can then have topological implications for the manifold. Thus we can derive topological obstructions to the existence of positive scalar curvature metrics (§3.4).

§3.1 *Dirac operators*

Before introducing the Dirac operator, a preliminary discussion about connections on spinor bundles is in order. (The following material is an outline of the key facts we will need. All the details behind these statements can be found in [LM; II.4].)

Given a choice of connection in the fibres of a spin bundle E - in particular a metric connection compatible with a choice of Riemannian metric in the fibres - we obtain an induced principal connection on the associated principal spin bundle P. Given a connection on P, there is a natural way of producing a connection, and thus a covariant derivative on any bundle associated to P. This applies in particular to Clifford and spinor bundles $Cl(E)$ and $S(E)$.

This covariant derivative on $Cl(E)$ has the property that Clifford multiplication acts as a derivation: $\nabla(\sigma \cdot \tau) = (\nabla\sigma) \cdot \tau + \sigma \cdot (\nabla\tau)$ for $\sigma, \tau \in \Gamma(Cl(E))$. For the spinor bundle, covariant differentiation is a derivation with respect to the module structure over $Cl(E)$: $\nabla(\sigma \cdot s) = (\nabla\sigma) \cdot s + \sigma \cdot (\nabla s)$ for $\sigma \in \Gamma(Cl(E))$ and $s \in \Gamma(S(E))$. As a consequence we see that ∇ preserves the splitting $Cl(E) = Cl^0(E) \oplus Cl^1(E)$.

The volume section $\omega \subset \Gamma(Cl(E))$ is *parallel* for our covariant derivative. Consequently, in dimensions where we have a volume section splitting of Clifford and spinor bundles, these splittings are also preserved by ∇.

We will assume from now on that every spinor bundle S is equipped with an inner product in the fibres having the property that Clifford multiplication by unit vectors is orthogonal, i.e. $\langle e \cdot s, e \cdot s' \rangle = \langle s, s' \rangle$, as well as a connection ∇ which is compatible with this Riemannian metric.

We will specialise to the following situation. Let X^n be a oriented Riemannian spin manifold with tangent bundle TX. Denote by $Cl(X)$ the Clifford bundle associated to TX, and let S be any spinor bundle associated to TX.

Definition 3.1.1. *The Dirac operator is the linear first order differential operator* $D : \Gamma(S) \to \Gamma(S)$ *defined by*

$$D(\sigma) = \sum_{i=1}^{n} e_i \cdot \nabla_{e_i}\sigma,$$

W. Tuschmann, D.J. Wraith, *Moduli Spaces of Riemannian Metrics*,
Oberwolfach Seminars 46, DOI 10.1007/978-3-0348-0948-1_3

where $\{e_i\}$ is any choice of orthonormal basis for T_xX at each point $x \in X$, and where $\Gamma(S)$ denotes the space of smooth sections of S.

It is easy to check that this definition is independent of the orthonormal bases used.

Let us consider the example when $X = S^1$. In this case we have $TS^1 \cong S^1 \times \mathbb{R}$, so

$$Cl(S^1) = S^1 \times Cl_1 = S^1 \times \mathbb{C}.$$

The structural group is SO(1), which is trivial. Thus the corresponding principal spin bundle could be either $S^1 \times \mathbb{Z}_2$ or S^1 (as a double cover of itself). Let us choose the first of these. The unique irreducible Clifford module for $\mathrm{Spin}(1) = \mathbb{Z}_2$ is $\mathbb{C}$, so as our (unique irreducible) spinor bundle we have the product $S^1 \times \mathbb{C}$ (which coincides with $Cl(S^1)$). A unit vector in TS^1 corresponds to $i \in \mathbb{C}$, therefore the Dirac operator in this case is $D = i\partial/\partial\theta$ (where θ is the angular parameter in S^1) acting on functions $f : S^1 \to \mathbb{C}$. Notice that $D^2 = -\partial^2/\partial\theta^2$, the Laplacian for functions on the circle.

For another elementary example, consider the Dirac operator on $\mathbb{R}^2$. The tangent bundle $T\mathbb{R}^2 = \mathbb{R}^2 \times \mathbb{R}^2$ is trivial, as is the corresponding Clifford bundle $Cl(\mathbb{R}^2) = \mathbb{R}^2 \times Cl_2 = \mathbb{R}^2 \times \mathbb{H}$, which in this case agrees with the spinor bundle $S(\mathbb{R}^2)$. Note that the Dirac operator on $\mathbb{R}^2$ acts on sections of this spinor bundle, which can be viewed as functions $\mathbb{R}^2 \to \mathbb{H}$. In terms of the standard basis $\{e_1, e_2\}$ for (each tangent space of) $\mathbb{R}^2$, we have $Cl_2 = \mathrm{Span}\{1, e_1, e_2, e_1 \cdot e_2\}$, and we can fix an isomorphism of Clifford algebras $Cl_2 \cong \mathbb{H}$ such that e_1 corresponds to i, e_2 to j and $e_1 \cdot e_2$ to k. We thus obtain

$$D = e_1 \cdot \nabla_{e_1} + e_2 \cdot \nabla_{e_2} = i\frac{\partial}{\partial x} + j\frac{\partial}{\partial y}.$$

This gives

$$D^2 = -\frac{\partial^2}{\partial x^2} - \frac{\partial^2}{\partial y^2} + ij\frac{\partial^2}{\partial x \partial y} + ji\frac{\partial^2}{\partial y \partial x} = -\frac{\partial^2}{\partial x^2} - \frac{\partial^2}{\partial y^2},$$

and so again we see that the Dirac operator squares to give the Laplacian.

It is important to note, however, that the above examples are not representative in the sense that typically a Dirac operator will *not* square to give a Laplacian. The second order term will always agree with that of a Laplacian (hence we might call D^2 an 'operator of Laplace type'), however lower order terms will in general differ.

Working out the details of the Dirac operator even in simple cases can be very difficult. Fortunately this will not concern us: what is of interest to us is the existence of Dirac operators, spinor bundles and related phenomena. From this alone we are able to derive powerful conclusions.

We will assume from now on that X is closed, that is, compact and without boundary. Consider the L^2-inner product on $\Gamma(S)$ given by

$$(s_1, s_2) := \int_X \langle s_1(x), s_2(x) \rangle \, dx$$

where $\langle\, , \rangle$ indicates the inner products in the fibres of S. With respect to this L^2-inner product we have the following consequence of the divergence theorem:

Lemma 3.1.2. *For any $s_1, s_2 \in \Gamma(S)$ we have $(D(s_1), s_2) = (s_1, D(s_2))$, that is, D is self-adjoint.*

An important feature of a Dirac operator is that it is an *elliptic* operator. This has many consequences, the most important of which for our purposes are the following facts.

Fact 1. *The kernel and cokernel of any Dirac operator are finite dimensional.*

Note that the cokernel of a linear map $f : V \to W$ is the quotient $W/\overline{\text{im} f}$. Fact 1 is striking as the domain and range of the Dirac operator is the infinite dimensional space $\Gamma(S)$. (See the 'technical point' at the end of the section for more on this.)

Self-adjointness together with ellipticity give

Fact 2. *([LM] III 5.8). Each eigenspace of D is finite-dimensional and consists of smooth sections. The eigenvalues are real, discrete and tend rapidly to infinity. Moreover $L^2(S)$ is the orthogonal direct sum of the eigenspaces.*

It is vital to note that requiring X to be a spin manifold and S to be a spinor bundle associated to TX is not necessary in the definition of the Dirac operator. In fact we merely need X to be an oriented Riemannian manifold, and S to be a bundle of left modules for $Cl(X)$ equipped with a Riemannian metric in its fibres and a connection as described at the start of this section. See [LM; page 114] for details. These more general forms of Dirac operators are important, and can be used for example to describe classical invariants such as the Euler characteristic and signature of a manifold (see [LM; II.6]). The way in which Dirac operators are associated to these and other invariants involves a concept called the 'index' of an operator. We will discuss this next.

§3.2 *The index*

The index of a linear operator is a concept which later on will be crucial for studying positive scalar curvature. We will motivate the concept by showing how it arises in elementary linear algebra. Given a linear map $f : V \to W$ between finite dimensional vector spaces, the dimension of the cokernel $\text{dimcoker} f = \text{dim} W - \text{dim im} f$. Recall that the Rank-Nullity Theorem for such maps says that $\text{dimker} f + \text{dim im} f = \text{dim} V$. We can re-write this as

$$\text{dimker} f - \text{dimcoker} f = \text{dim} V - \text{dim} W.$$

Thus the quantity $\text{dimker} f - \text{dimcoker} f$ is constant for all such linear maps.

With a view to studying linear operators on infinite dimensional spaces, we make the following:

Definition 3.2.1. *The index of a linear map $f : V \to W$ is the quantity $\text{ind} f = \text{dimker} f - \text{dimcoker} f$.*

It is clear that this will not make sense in general for operators on infinite dimensional spaces, however in many important cases the index is finite.

Let us illustrate this concept by giving a very simple example of a linear operator between infinite dimensional spaces with finite dimensional kernel, cokernel, and therefore index. Consider the Hilbert space of real sequences ℓ^2, and the 'right-shift' operator $R : (a_1, a_2, a_3, ...) \mapsto (0, a_1, a_2, a_3, ...)$. Clearly the kernel of this is trivial, and the cokernel is isomorphic to $\mathbb{R}$, thus the index is -1. Similarly the left-shift operator $L : (a_1, a_2, a_3, ...) \mapsto (a_2, a_3, a_4, ...)$ has index $+1$. Compositions of right or left-shift operators then provide examples of operators with index equal to any given integer.

There is a problem with studying the index of a Dirac operator, however. As a Dirac operator is self adjoint, this means that the kernel and cokernel are isomorphic. As a result we deduce that *the index of the Dirac operator vanishes.*

Recall, however, that in dimensions 0 mod 4 in the real case and in all even dimensions in the complex case, the spinor bundles split as $S = S^+ \oplus S^-$, and Clifford multiplication by a vector $e \in TX$ on these subspaces gives a map $S^\pm \to S^\mp$. As a consequence we see that the Dirac operator splits as $D = D^+ \oplus D^-$ with $D^+ : S^+ \to S^-$ and $D^- : S^- \to S^+$. Now $D^\pm$ are still elliptic differential operators, *but are not self-adjoint.* In fact D^+ is the adjoint of D^-, and D^- is the adjoint of D^+. When we talk about the index of a Dirac operator, we generally mean the index $\mathrm{ind}D^+$.

It turns out that $\mathrm{ind}D^+$ contains important geometric and topological information. This index is quite stable under perturbations, and consequently can be described in purely topological terms. This description is one of the major achievements of twentieth century mathematics, and we will discuss aspects of this in the next section.

Technical point. The fact that the kernel and cokernel of the Dirac operators D and D^+ are finite involves more than one might at first imagine. As suggested above, this finiteness arises from the theory of elliptic differential operators. However in order to deduce this we need to work with a *bounded* operator, and unfortunately the Dirac operator is not bounded. On the other hand there is a natural extension to an operator between Sobolev spaces which *is* bounded. Ellipticity of this extension then guarantees that it has finite index. Moreover the index is independent of how this extension is performed, and the index of the original Dirac operator is then taken to be the index of any of its extensions. For more details see [LM, III.5], especially Theorem 5.2(ii). We also need the Sobolev extensions and Hilbert space properties of Sobolev spaces (see for example [Ru; chapter 12]) to deduce that the kernel and cokernel of the Dirac operator are isomorphic.

§3.3 *The Atiyah-Singer index theorem*

Let X^{2n} be an even dimensional, closed, oriented spin manifold. Let $\mathbb{S}$ be the irreducible complex spinor bundle over X, and consider the Dirac operator $D^+ : \Gamma(\mathbb{S}^+) \to \Gamma(\mathbb{S}^-)$. We will call this operator the *Atiyah-Singer Dirac operator.* (Note that in [LM] this operator is called the 'Atiyah-Singer $\hat{A}$-operator', with the term 'Atiyah-Singer operator' being used for the Dirac operator on spinor bundles.)

Theorem 3.3.1. *([AS3]) The index of the Atiyah-Singer Dirac operator D^+ is given by the following formula*

$$indD^+ = \hat{A}(X)$$

where $\hat{A}(X)$ denotes the $\hat{A}$-genus of X.

As X is oriented, we have a fundamental homology class $[X] \in H_{2n}(X;\mathbb{Q})$. The $\hat{A}$-genus of X is defined by $\langle \hat{A}(TX), [X]\rangle$ where $\langle\,,\rangle$ denotes here the evaluation pairing of cohomology on homology, and $\hat{A}(TX)$ is the $\hat{A}$-polynomial of the tangent bundle TX. This polynomial is a polynomial in the rational Pontrjagin classes $p_i \in H^{4i}(X;\mathbb{Q})$ of the tangent bundle. Thus if the dimension of X is equivalent to 2 mod 4, the $\hat{A}$-genus and hence the index is zero. However if the dimension of X is a multiple of 4, both the $\hat{A}$-genus and the index in many cases are non-zero. We therefore focus exclusively on these dimensions.

The $\hat{A}$-polynomial can be defined for any real vector bundle E over a manifold M (not just the tangent bundle and not just over spin manifolds, and without any restriction on dimension). It takes the form

$$\hat{A}(E) = 1 + \hat{A}_1(p_1(E)) + \hat{A}_2(p_1(E), p_2(E)) + \hat{A}_3(p_1(E), p_2(E), p_3(E)) + \ldots \in H^*(M;\mathbb{Q}).$$

Here, each $\hat{A}_k(p_1(E), \ldots, p_k(E))$ is a homogeneous degree k polynomial, which is an element of $H^{4k}(M;\mathbb{Q})$. As noted above, if M is an oriented manifold we can evaluate the $\hat{A}$-polynomial on the fundamental homology class to obtain the $\hat{A}$-genus of M. Clearly this will be zero unless M has dimension $4n$, and in this case only the degree $4n$ term $\hat{A}_n(p_1(E), \ldots, p_n(E))$ will play a role. These polynomials, which are ultimately derived from the properties of a certain real valued function, take the following form for small k:

$$\hat{A}_1(p_1) = -\frac{1}{24}p_1;$$

$$\hat{A}_2(p_1, p_2) = \frac{1}{2^7 \cdot 3^2 \cdot 5}(-4p_2 + 7p_1^2);$$

$$\hat{A}_3(p_1, p_2, p_3) = \frac{1}{2^{10} \cdot 3^3 \cdot 35}(16p_3 - 44p_2p_1 + 31p_1^3).$$

In general, the $\hat{A}$-genus of a manifold will not be an integer. For example $\hat{A}(\mathbb{C}P^2) = -1/8$, and more generally $\hat{A}(\mathbb{C}P^{2k}) = (-1)^k 2^{-4k}(2k)!/(k!)^2$. However the index theorem shows that *for a spin manifold, the $\hat{A}$-genus is always an integer.* For more details about the $\hat{A}$-genus and the definition of the $\hat{A}$-polynomial see [LM; III.11].

We can also think of the $\hat{A}$-genus of a Riemannian manifold $(X^{4n}; g)$ in terms of deRham cohomology. Corresponding to the (real) Pontrjagin classes are certain 'Pontrjagin forms' $p_i = p_i(X;g) \in \Omega^{4i}(X)$, where $\Omega^*(X)$ is the exterior algebra of differential forms on the manifold X. These forms, which are built from the curvature form (see [Mo; §5.4]), are metric-dependent, however the cohomology classes they represent are not. We can form 'polynomials' $\hat{A}_i$ from these forms exactly as before, simply replacing the cup product of cohomology classes with the exterior product (wedge) of differential forms. The Atiyah-Singer index theorem then reads

$$\mathrm{ind} D^+ = \int_X \hat{A}_n(p_1, \ldots, p_n).$$

The $\hat{A}$ genus has an important multiplicative property: $\hat{A}(M \times N) = \hat{A}(M)\hat{A}(N)$. For example $\hat{A}(\mathbb{C}P^2 \times \mathbb{C}P^2) = +1/64$. (However this property does not extend to bundles - see [HSS].)

More generally, the $\hat{A}$-genus defines a ring homomorphism

$$\hat{A} : \Omega_*^{SO} \to \mathbb{Q},$$

where Ω_*^{SO} denotes the oriented bordism ring. As a consequence we have $\hat{A}(M\sharp N) = \hat{A}(M) + \hat{A}(N)$. For example $\hat{A}(\mathbb{C}P^2\sharp(S^2 \times S^2)) = -1/8 + \hat{A}(S^2 \times S^2) = -1/8 + 0 = -1/8$. (Elements of Ω_*^{SO} are equivalence classes of oriented manifolds, where the equivalence relation is given by $M^n \sim N^n$ if and only if there is an oriented manifold W^{n+1} with boundary $M \coprod N$, for which the orientation agrees with those on M and N. The additive operation in Ω_*^{SO} is disjoint union, and the multiplicative operation is the Cartesian product. Note that the disjoint union $M \coprod N$ of two oriented manifolds is always oriented bordant to the connected sum $-(M\sharp N)$, so we could equally take the additive operation to be connected sum. If we consider spin structures in place of orientations, we can define the spin bordism ring Ω_*^{spin} in an analogous fashion.)

For spin manifolds there is a ring homomorphism

$$\alpha : \Omega_*^{spin} \to KO^{-*}(pt)$$

which in some sense generalizes the $\hat{A}$-genus. We will see that this plays an important role in positive scalar curvature geometry. For more details about K-theory, the groups $KO^{-*}(pt)$ and the definition of α, see Appendix A. Like the $\hat{A}$-genus, the α-invariant can also obtained as an index, but a generalized notion of index for a special variant of the Dirac operator. This variant Dirac operator is a so-called Cl_k-linear Dirac operator: this is a Dirac operator on a spinor bundle over X carrying a right-action of Cl_k which is parallel and commutes with multiplication from $Cl(X)$. (As in the case of the basic Dirac operator, we can extend this definition to bundles of left $Cl(X)$-modules over X, irrespective or whether X is spin or not, provided the fibres are equipped with a suitable metric and metric connection.) One can define an index for such operators in $KO^{-k}(pt)$. (See Appendix A for details.)

The basic case of a Cl_k-linear Dirac operator is as follows: let $P_{Spin}(X)$ be the principal Spin(n)-bundle associated to (the tangent bundle of) a spin manifold X^n, and set $S(X) := P_{Spin}(X) \times_\ell Cl_n$, where ℓ denotes the usual left multiplication of Cl_n by elements of Spin(n). It is not difficult to see that $S(X)$ admits a right Cl_n-action with the required properties (see [LM; page 139]). The splitting $Cl_n = Cl_n^0 \oplus Cl_n^1$ results in a bundle splitting $S(X) = S^0(X) \oplus S^1(X)$, which gives a $\mathbb{Z}_2$-grading over both the usual left $Cl(X)$-action and also the right Cl_n-action. We obtain a restricted Dirac operator

$$D^0 : \Gamma(S^0(X)) \to \Gamma(S^1(X))$$

called the Cl_n-linear Atiyah-Singer operator. The index of this operator is $\alpha(X)$. Note that there is an isomorphism $KO^{-4m} \cong \mathbb{Z}$ for all m under which we have $\alpha(M^{4m}) = \hat{A}(M)$.

As an example, there are exotic spheres Σ^{8k+i} for all $k \geq 1$ and $i = 1, 2$ which satisfy

$$\alpha(\Sigma^{8k+i}) \neq 0 \in KO^{-(8k+i)}(pt) \cong \mathbb{Z}_2.$$

We will meet these objects again later.

We can apply the above arguments to study Dirac operators in different contexts and derive different index results. A particularly famous example involves starting with a closed oriented manifold X^{4n} (not necessarily spin) and considering the Clifford bundle $Cl(X)$. In these dimensions we can split this bundle into sub-bundles $Cl^+(X)$ and $Cl^-(X)$ using the volume section $(-1)^n\omega$ (precisely, split $Cl(X)$ into the ± 1-eigenspaces for left multiplication by $(-1)^n\omega$). We can define a Dirac operator on $\Gamma(Cl(X))$ just as before, and this splits as $D^\pm : \Gamma(Cl^\pm(X)) \to \Gamma(Cl^\mp(X))$. The index of this operator gives a formula analogous to that in Theorem 3.3.1:

$$\mathrm{ind} D^+ = L(X),$$

where $L(X)$ is the so-called L-genus of X. Like the $\hat{A}$-genus, the L-genus of an oriented manifold is obtained using certain polynomial expressions in the Pontrjagin classes of the tangent bundle. These 'L-polynomials' can be defined for any vector bundle E in a similar fashion to the $\hat{A}$-polynomials:

$$L(E) = 1 + L_1(p_1) + L_2(p_1, p_2) + L_3(p_1, p_2, p_3) + \ldots$$

We then set $L(X) := \langle L(TX), [X] \rangle$, i.e. we evaluate the L-polynomial of the tangent bundle on the fundamental homology class of X. The first few of these polynomials are

$$L_1(p_1) = \frac{1}{3} p_1;$$

$$L_2(p_1, p_2) = \frac{1}{45}(7p_2 - p_1^2);$$

$$L_3(p_1, p_2, p_3) = \frac{1}{945}(62p_3 - 13p_1p_2 + 2p_1^3).$$

Recall that the signature of a $4n$-dimensional manifold is defined as follows: the cup product on $H^{2n}(X;\mathbb{Z})$ is a symmetric bilinear form. The signature of X is defined to be the signature of this form, that is, the number of positive eigenvalues minus the number of negative eigenvalues. The Hirzebruch signature theorem (see for example [MS; §19]) asserts that the signature $\sigma(X)$ of X is given by $\sigma(X) = L(X)$. Thus for this Dirac operator we have $\mathrm{ind} D^+ = \sigma(X)$, and for this reason we call D^+ the *signature operator*. Note that for a Riemannian manifold we could also define the L-genus via deRham cohomology using 'polynomials' of Pontrjagin forms on the manifold, and thus express the signature as an integral of the top dimensional form. For more details about the L-genus and signature theorem, see for example [MS, §19].

§3.4 *The index and positive scalar curvature*

Given a Riemannian vector bundle E over a manifold X^n with connection ∇ in the fibres, we define the second covariant derivative ∇^2 by

$$\nabla^2_{V,W}\phi = \nabla_V \nabla_W \phi - \nabla_{\nabla_V W}\phi$$

where V and W are tangent vector fields to X, and $\phi \in \Gamma(E)$. Notice that ∇^2 is tensorial in all its arguments. The *connection Laplacian* $\nabla^*\nabla$ is then defined by

$$\nabla^*\nabla\phi := -\sum_{i=1}^{n} \nabla^2_{e_i,e_i}\phi,$$

where $\{e_i\}$ is a local orthonormal tangent frame field.

The basic result linking Dirac operators to the scalar curvature is the following classical result, due to Lichnerowicz:

Theorem 3.4.1. *Let X be a Riemannian spin manifold and S any bundle of spinors over X with its canonical Riemannian connection. If D is the (full) Dirac operator we have*

$$D^2 = \nabla^*\nabla + \frac{1}{4}\kappa,$$

where κ denotes the scalar curvature of X.

It is not difficult to show that $\nabla^*\nabla$ is L^2-self adjoint on closed manifolds in exactly the same way as for the Dirac operator. As a consequence of self-adjointness, we see that $\ker D^2 = \ker D$ and $\ker\nabla^*\nabla = \ker\nabla$. The argument is this: $D^2(s) = 0$ implies $(D^2(s), s) = 0$. But by self-adjointness this means $(D(s), D(s)) = 0$ and hence $s \in \ker D$. The converse is trivial. Similarly for $\nabla^*\nabla$. We also have $(\nabla^*\nabla s, s') = (\nabla s, \nabla s')$. Combining this last observation with the Lichnerowicz formula, we deduce the following

Corollary 3.4.2. *If the scalar curvature of a closed spin manifold X is positive, then $\ker D$ is trivial.*

Proof. Let $s \in \ker D$. Then integrating over X gives

$$(D^2(s), s) = (\nabla^*\nabla(s), s) + \frac{1}{4}\int_X \kappa\langle s, s\rangle\, dx$$

where $\langle\, ,\, \rangle$ is the inner product in the fibres, and so

$$0 = |\nabla(s)|^2_{L^2} + \frac{1}{4}\int_X \kappa\langle s, s\rangle\, dx.$$

Now if $\kappa > 0$ we have a contradiction unless s is the zero section. □

Sections in $\ker D$ are called *harmonic spinors.* Hence the above Corollary can be re-phrased as: on a closed spin manifold with positive scalar curvature, there are no non-trivial harmonic spinors. Notice also that this is true *independent of the spinor bundle used.*

In situations where the Dirac operator splits as $D = D^+ \oplus D^-$, if $\ker D = 0$, we deduce that both $\ker D^+ = 0$ and $\ker D^- \cong \operatorname{coker} D^+ = 0$, and so

$$\operatorname{ind} D^+ = \dim\ker D^+ - \dim\operatorname{coker} D^+ = 0.$$

If we combine this observation with the Atiyah-Singer index theorem we obtain the following:

Theorem 3.4.3. *If X^{4n} is a closed spin manifold which admits a positive scalar curvature metric, then $\hat{A}(X) = 0$.*

Thus the $\hat{A}$-genus is a topological obstruction to positive scalar curvature in dimensions $4n$. In fact we can say more:

Theorem 3.4.4. *([St]) A closed simply connected manifold X^n of dimension at least 5 admits a positive scalar curvature metric if and only if $\alpha(X) = 0 \in KO^{-n}(pt)$.*

For example the exotic spheres Σ^{8k+i} mentioned above with $k \geq 1$ and $i = 1, 2$ which have non-zero α-invariant cannot admit any metric with positive scalar curvature.

4. Early results about the space of positive scalar curvature metrics

In this chapter we discuss classical results of Hitchin and Carr. The results of Hitchin were the first to show that the topology of the space of positive scalar curvature metrics could be non-trivial. Specifically, he showed that in certain dimensions there are closed spin manifolds for which π_0 and π_1 of the space of positive scalar curvature metrics are non-trivial, by exploiting properties of diffeomorphism groups. However these results do not indicate the extent to which the topology can be non-trivial. Taking a very different approach, the work of Carr showed that the space of positive scalar curvature metrics on the spheres S^{4n-1} has in fact infinitely many path components. We conclude the chapter by mentioning recent work of Crowley and Schick which extends Hitchin's results.

§4.1 *The work of Hitchin*

There are three main ingredients in Hitchin's arguments: a filtration on the group of homotopy spheres in each dimension; a construction of certain bundles with special properties, and finally some results from index theory. We treat each of these separately.

§4.1.1 *Gromoll groups*

Let Θ_{n+1} denote the group of oriented diffeomorphism classes of homotopy $(n+1)$-spheres $(n > 5)$. Any homotopy $(n+1)$-sphere Σ^{n+1} can be viewed as the union of two discs D^{n+1} glued together on their boundary by a diffeomorphism $\phi \in \mathrm{Diff}(S^n)$. Conversely any such gluing produces a homotopy sphere. Thus we have a map $\mathrm{Diff}(S^n) \to \Theta_{n+1}$, and indeed a surjective homomorphism

$$\pi_0\mathrm{Diff}(S^n) \to \Theta_{n+1}.$$

We can use this view of homotopy spheres to put a filtration on each group Θ_{n+1}, called the *Gromoll filtration.*

Firstly we replace $\mathrm{Diff}(S^n)$ by the group $\mathrm{Diff}(D^n, S^{n-1})$, the group of orientation preserving diffeomorphisms of D^n which restrict to the identity near the boundary. There is clearly a map $\mathrm{Diff}(D^n, S^{n-1}) \to \mathrm{Diff}(S^n)$ given by including D^n into S^n as the upper hemisphere say, and extending by the identity to a diffeomorphism of S^n. In turn this induces a map

$$\iota_n : \pi_0\mathrm{Diff}(D^n, S^{n-1}) \to \pi_0\mathrm{Diff}(S^n).$$

Composing ι_n with the homomorphism $\pi_0\mathrm{Diff}(S^n) \to \Theta_{n+1}$ above gives a map

$$\tau_n : \pi_0\mathrm{Diff}(D^n, S^{n-1}) \to \Theta_{n+1}.$$

It turns out ([Ce2], [Sm]) that τ_n is in fact an *isomorphism.* Thus it suffices to work with the group $\mathrm{Diff}(D^n, S^{n-1})$. We also have maps

$$\lambda^n_{i,j} : \pi_j\mathrm{Diff}(D^{n-j}, S^{n-j-1}) \to \pi_{j-i}\mathrm{Diff}(D^{n-j+i}, S^{n-j+i-1})$$

W. Tuschmann, D.J. Wraith, *Moduli Spaces of Riemannian Metrics*,
Oberwolfach Seminars 46, DOI 10.1007/978-3-0348-0948-1_4

for $0 < i \leq j$ defined as follows. We view discs as unit cubes, for example $D^n = [0,1]^n$, and homotopy groups $\pi_j \mathrm{Diff}(D^{n-j}, S^{n-j-1})$ as relative homotopy groups

$$[([0,1]^j, \partial[0,1]^j), (\mathrm{Diff}(D^{n-j}, S^{n-j-1}), \{id\})].$$

We can then represent an element $[\phi] \in \pi_j \mathrm{Diff}(D^{n-j}, S^{n-j-1})$ by a map

$$\phi : [0,1]^j \to \mathrm{Diff}([0,1]^{n-j}, \partial[0,1]^{n-j})$$

where ϕ is the identity near the boundary of $[0,1]^j$ and the image of any point in $[0,1]^j$ is a diffeomorphism of $[0,1]^{n-j}$ which restricts to the identity near the boundary of $[0,1]^{n-j}$. It will be convenient below to first define $\lambda_{i,j}^n$ as a map acting on ϕ, and then later to be the corresponding induced map on homotopy classes.

We define $\lambda_{i,j}^n(\phi)$ by re-writing the above expression for ϕ by splitting $[0,1]^j$ as a product $[0,1]^i \times [0,1]^{j-i}$:

$$\lambda_{i,j}^n(\phi) : [0,1]^{j-i} \to \mathrm{Diff}([0,1]^{n-j} \times [0,1]^i, \partial)$$

where $\lambda_{i,j}^n(\phi)(z)(x,y) := (\phi(y,z)(x), y)$ with $x \in [0,1]^{n-j}$, $y \in [0,1]^i$, $z \in [0,1]^{j-i}$, and where $\partial = \partial([0,1]^{n-j} \times [0,1]^i)$.

As a special case of this construction suppose that $i = j$. The domain of $\lambda_{j,j}(\phi)$ is just a one-point space, so in effect we can view $\lambda_{j,j}$ as the map $(x,y) \mapsto (\phi(y)(x), y)$ in $\mathrm{Diff}(D^n, S^{n-1})$. As this map is the identity on the $[0,1]^j$ factor of $[0,1]^n = [0,1]^{n-j} \times [0,1]^j$, it is clear that the collection of diffeomorphisms in the image of $\lambda_{j,j}$ will form a decreasing family of nested sets as j increases.

The induced map on homotopy

$$\lambda_{j,j} : \pi_j \mathrm{Diff}(D^{n-j}, S^{n-j-1}) \to \pi_0 \mathrm{Diff}(D^n, S^{n-1})$$

can be combined with the isomorphism $\tau_n : \pi_0 \mathrm{Diff}(D^n, S^{n-1}) \cong \Theta_{n+1}$ to give a homomorphism

$$\lambda_j : \pi_j \mathrm{Diff}(D^{n-j}, S^{n-j-1}) \to \Theta_{n+1}.$$

Notice what we are doing here: we are in effect taking a diffeomorphism of D^n produced by $\lambda_{j,j}$ and extending it by the identity map to a diffeomorphism on all of S^n. This is then used as a map for gluing the boundaries of two copies of D^{n+1} to produce a homotopy sphere Σ^{n+1}. The *Gromoll group* Γ_{j+1}^{n+1} is the subgroup of Θ_{n+1} defined as the image of the map λ_j. By the remark above, we should expect (and indeed we prove below) that for all j we have $\Gamma_{j+1}^{n+1} \subset \Gamma_j^{n+1}$. We can interpret this as saying that the larger the j, the more topologically simple the homotopy sphere in the image of λ_j.

An alternative way of viewing the Gromoll group Γ_{j+1}^{n+1} is as follows: view an element $\sigma \in \pi_j \mathrm{Diff}(S^{n-j})$ as a map $\sigma : D^j \to \mathrm{Diff}(S^{n-j})$ which is the identity near the boundary of D^j. This then gives a diffeomorphism $\sigma' : D^j \times S^{n-j} \to D^j \times S^{n-j}$ by setting $\sigma'(x,y) = (x, \sigma(x)y)$. Using this we can create a diffeomorphism of S^n by viewing S^n as $D^j \times S^{n-j} \cup S^{j-1} \times D^{n-j+1}$ (i.e. as the boundary of $D^j \times D^{n-j+1}$) and extending σ' to all of S^n by the identity on $S^{j-1} \times D^{n-j+1}$. By associating this diffeomorphism with the corresponding homotopy sphere, we obtain a map (actually a homomorphism) $\lambda_j' : \pi_j \mathrm{Diff}(S^{n-j}) \to \Theta_{n+1}$. This is essentially the same map as λ_j above, and in particular has the same image, namely Γ_{j+1}^{n+1}.

Lemma 4.1.1.1. *For all j we have $\Gamma^{n+1}_{j+1} \subset \Gamma^{n+1}_{j}$.*

Proof. The result follows easily from the fact that the map λ_j satisfies

$$\lambda_j = \lambda_{j-1} \circ \lambda_{j,1},$$

that is, the map λ_j factors through the group $\pi_{j-1}\mathrm{Diff}(D^{n-j+1}, S^{n-j})$. Let $a \in [0,1]^{n-j}$, $b \in [0,1]^1$ and $c \in [0,1]^{j-1}$: we can then view (b,c) as an element of $[0,1]^j$, for example, and $(a,b,c) \in [0,1]^n$. For $\phi \in \pi_j\mathrm{Diff}([0,1]^{n-j}, \partial[0,1]^{n-j})$, set $\psi := \lambda_{j,1}(\phi)$. Then

$$\begin{aligned}\lambda_{j-1,j-1}(\psi)(a,b,c) &= (\psi(c)(a,b),c)\\ &= (\lambda_{j,1}(\phi)(c)(a,b),c)\\ &= (\phi(b,c)(a),b,c).\end{aligned}$$

We compare this with $\lambda_{j,j}(\phi)(a,b,c) = (\phi(b,c)(a),(b,c))$, which clearly agrees with the above when we identify $[0,1]^1 \times [0,1]^{j-1}$ with $[0,1]^j$. □

Corollary 4.1.1.2. *There is a filtration*

$$0 = \Gamma^{n+1}_{n-1} \subset \Gamma^{n+1}_{n-2} \subset \Gamma^{n+1}_{n-3} \subset \cdots \subset \Gamma^{n+1}_{2} \subset \Gamma^{n+1}_{1} = \Theta_{n+1}.$$

Remark 4.1.1.3. *In fact we can say more than this. By a result of Cerf [Ce2] (as pointed out in [ABK]) we have $\Gamma^{n+1}_2 = \Theta_{n+1}$. Moreover by a result of Hatcher [Ha] we have $\Gamma^{n+1}_{n-2} = 0$, so we actually have the filtration*

$$0 = \Gamma^{n+1}_{n-2} \subset \Gamma^{n+1}_{n-3} \subset \cdots \subset \Gamma^{n+1}_{3} \subset \Gamma^{n+1}_{2} = \Theta_{n+1}.$$

§4.1.2 *Hitchin bundles*

Given a closed manifold X^n and an element $[\phi] \in \pi_j\mathrm{Diff}(D^n, S^{n-1})$, we show how to construct a bundle over S^{j+1} with fibre X. By choosing an embedded disc $D^n_X \subset X$, we can regard ϕ as acting on X, and thus ϕ determines an element $\phi_X \in \pi_j\mathrm{Diff}(X)$. We therefore obtain a diffeomorphism $\Phi : X \times S^j \to X \times S^j$ such that $\Phi(x,s) := (\phi_X(s)(x), s)$. We can then construct a bundle over S^{j+1} with fibre X by letting Φ act as clutching function: $(X \times D^{j+1}) \cup_\Phi (X \times D^{j+1})$. Call the total space of this bundle Z.

It is useful to compare Z with the trivial bundle $X \times S^{j+1} = (X \times D^{j+1}) \cup_{id} (X \times D^{j+1})$ by comparing clutching functions. By a small homotopic adjustment if necessary, we can assume the clutching function Φ is the identity map on $X \times U$ for some open neighbourhood U of a base point in S^j. Restrict Φ to the part of its domain where it is *not* the identity, that is, to $D^n_X \times D^j$ where we have set $D^j = S^j \setminus U$. We can view this restriction as a map

$$\Phi_{D^n_X \times D^j} : [0,1]^n \times [0,1]^j \to [0,1]^n \times [0,1]^j$$

given by

$$\Phi_{D^n_X \times D^j}(x,y) = (\phi_X(y)(x), y).$$

But notice that this agrees with the map $\lambda_{j,j}(\phi)$ when we identify D^n with D^n_X.

Let us widen our perspective now and consider the whole bundles Z and $X \times S^{j+1}$. From the above arguments these spaces differ in the following way: remove a disc $D^{n+j+1} = D^n_X \times (D^j \times [0, \epsilon])$ from the trivial bundle, where the $[0, \epsilon]$ factor is an interval for the normal parameter to the equator of S^{j+1}, with 0 corresponding to the equator. To create Z from this, we must glue back the disc D^{n+j+1} via a diffeomorphism of the boundaries which is the identity on all sides except that along the equator, where it agrees with $\Phi_{D^n_X \times D^j}$. Thus this boundary diffeomorphism $S^{n+j} \to S^{n+j}$ is produced by extending $\Phi_{D^n_X \times D^j}$ by the identity over the remaining part of S^{n+j}. Comparing this gluing with the way in which the map λ_j is used to glue discs to form a homotopy sphere, we see that Z is simply $(X \times S^{j+1})X\sharp\Sigma$ where $\Sigma \in \Theta_{n+j+1}$ is the homotopy sphere defined by $\lambda_j(\phi)$.

The importance of this construction is that it creates examples of manifolds with non-trivial α-invariant (see §3.3). Recall that α is a ring homomorphism with domain the spin bordism ring. Thus if X is a spin manifold so is Z (as up to connected sum with a homotopy sphere Z is just a product of X and a sphere, thus $w_2(Z) = 0$ if and only if $w_2(X) = 0$), and $\alpha(Z) = \alpha(X)\alpha(S^{j+1}) + \alpha(\Sigma)$. As S^{j+1} admits positive scalar curvature we have $\alpha(S^{j+1}) = 0$, and hence $\alpha(Z) = \alpha(\Sigma)$. This is interesting as there are 'bad' exotic spheres in dimensions 1 and 2 modulo 8 with non-zero α-invariant (see Appendix A), so these spheres cannot admit any metric of positive scalar curvature (see §3.4). Thus any Z for which the corresponding exotic sphere is 'bad' must also have non-zero α-invariant, and therefore cannot admit positive scalar curvature.

To construct such Z we use the Gromoll filtration. In particular recall that $\Gamma^{n+1}_2 = \Gamma^{n+1}_1 = \Theta_{n+1}$. This means that if there are 'bad' exotic spheres in dimension $n+1$ or $n+2$ then there are elements in $\pi_0(\mathrm{Diff}(D^n, S^{n-1})$ respectively $\pi_1(\mathrm{Diff}(D^n, S^{n-1})$ which produce these spheres after applying the maps λ_0 respectively λ_1. The corresponding bundles then take the form $(X^n \times S^1)\sharp\Sigma^{n+1}$ and $(X^n \times S^2)\sharp\Sigma^{n+2}$. Of course these can also be described as X-bundles over S^2 respectively S^1. We obtain:

Theorem 4.1.2.1. *([Hi]) For any closed spin manifold X there are bundles $X^n \to Z \to S^2$ in dimensions congruent to 0 and 1 mod 8, and bundles $X^n \to Z \to S^1$ in dimensions congruent to 0 and 7 mod 8, for which the total space satisfies $\alpha(Z) \neq 0$.*

§4.1.3 *Spaces of metrics*

Now suppose that X^k is a closed spin manifold which admits a positive scalar curvature metric g. Denoting the space of positive scalar curvature metrics on X by $\mathcal{R}^+_{scal}(X)$, there is a map $T : \mathrm{Diff}(X) \to \mathcal{R}^+_{scal}(X)$ defined via pull back, $T(h) := h^*g$, and an induced map on homotopy $T_* : \pi_{n-1}\mathrm{Diff}(X) \to \pi_{n-1}\mathcal{R}^+_{scal}(X)$.

Given an element $\phi \in \pi_{n-1}(\mathrm{Diff}(X))$ we can construct a bundle with fibre X and base S^n in the usual way using any representative of the homotopy class ϕ as clutching function. We introduce metrics into the fibres of this bundle as follows. On one 'half' $D^n \times X$ we take the metric g in each of the fibres. In the other half we take the family of pull-back metrics $\phi(u)^*g$, $u \in \partial D^n$, in the fibres along the boundary S^{n-1}, and extend in any way to a smooth family of fibre metrics over the interior of the disc.

Now consider the family of Cl_k-linear Dirac operators (§3.3) in the fibres of this bundle. As the scalar curvature along the equator and in the 'southern hemisphere' say is positive, by the theorem of Lichnerowicz (§2.4) the operators on these fibres are invertible. This invertibility means that the operators in a neighbourhood of the closed southern hemisphere can be deformed in an essentially unique way to be constant (in the sense that the operators do not vary from point to point), see [LM; page 208]. In particular this deformation can be performed *without changing the index*. Thus we lose nothing if we focus attention on the northern hemisphere. We therefore have a family of Cl_k-linear operators in the fibres over a disc D^n which are invertible at each point of the boundary and constant along the boundary. There is a concept of index for such a family, which is an element of $KO^{-k}(D^n, S^{n-1}) \cong KO^{-k-n}(pt)$, and this can be shown to be independent of how the metrics on the fibres over the boundary of the northern hemisphere are extended over the interior fibres. We obtain a composite map

$$\pi_{n-1}\text{Diff}(X) \to \pi_{n-1}\mathcal{R}^+_{scal}(X) \to KO^{-k-n}(pt).$$

By [AS4], [AS5] and [Hi; Prop. 4.2], this index coincides with the α-invariant of the total space when the total space is a spin manifold.

If we consider again the examples Z of §4.1.2 with non-zero α-invariant, we observe that these are just formed from two copies of $X \times D^m$ for $m = 1, 2$ using some map ϕ as above. Thus the index of the family of operators with the fibre metrics chosen above is non-zero for these examples, and so $\pi_{n-1}\mathcal{R}^+_{scal}(X)$ must be non-trivial. We obtain

Theorem 4.1.3.1. *([Hi]). If X is a closed spin manifold which admits a positive scalar curvature metric, then*
(i) $\pi_0\mathcal{R}^+_{scal}(X) \neq 0$ when $\dim X \equiv 0, 1 \bmod 8$ (provided this dimension is ≥ 9);
(ii) $\pi_1\mathcal{R}^+_{scal}(X) \neq 0$ when $\dim X \equiv 7, 0 \bmod 8$ (provided this dimension is ≥ 7).

Note that these results do not survive to the moduli space, as they are based on the action of the diffeomorphism group of X.

§4.2 *The work of Carr*

Hitchin showed that in certain circumstances the space of positive scalar curvature metrics is not connected, however his results say nothing about the number of path components which might occur. A few years after the Hitchin paper, R. Carr ([Ca]) showed that the space of positive scalar curvature metrics on S^{4n-1} has *infinitely* many components for each $n \geq 2$. Carr's approach is somewhat different from Hitchin's. It does not use any properties of diffeomorphism groups, rather it uses a certain property of positive scalar curvature metrics together with the $\hat{A}$-genus obstruction. The ideas are for the most part not tied to the spheres S^{4n-1}, meaning that his approach has applicability beyond the examples he studies.

There are various ingredients in Carr's argument. The first of these that we consider is the 'plumbing' construction.

§4.2.1 *Plumbing*

Plumbing is a construction by which disc bundles can be glued together to create a new manifold with boundary. The idea is as follows. (We refer the reader to [Br; § V] for more details.) Consider two disc bundles $D^n \hookrightarrow E_1 \to M^m$ and $D^m \hookrightarrow E_2 \to N^n$, and for each bundle choose a local trivialisation over a disc $D^m_M \subset M$ respectively $D^n_N \subset N$ in the base manifolds. Choose diffeomorphisms $\iota_M : D^m \to D^m_M$ and $\iota_N : D^n \to D^n_N$, and using the diffeomorphism $(\iota_M, \iota_N^{-1}) : D^m \times D^n_N \to D^m_M \times D^n$ in conjunction with the trivialisations, we glue the bundles E_1 and E_2 along the locally trivial neighbourhoods. Notice that this identification glues the base disc of one neighbourhood with the fibre disc of the other. The resulting object is the so-called plumbing of the disc bundles, and can be made differentiable by straightening out the angles. There are choices to be made in selecting the diffeomorphisms ι_M and ι_N, and these choices can influence the topology of the manifold produced. However we will always make the following assumptions which will mean the diffeomorphism type of the resulting manifold is unambiguous: we will assume that the fibres, base and total space for each bundle are compatibly oriented, and that the diffeomorphisms ι_M and ι_N are both orientation preserving. Notice that we can plumb any number of disc bundles together provided the fibre and disc dimensions correspond appropriately, or even plumb a bundle to itself.

The plumbing construction provides a way of explicitly realizing manifolds of dimension $4n$ with prescribed intersection form on middle dimension homology and thus prescribed signature. Using some algebraic topology this allows us to detect situations where the boundary is a homotopy sphere. The homotopy spheres which arise as the boundary of plumbed manifolds include all those which bound a parallelisable manifold (see [Wr1]) plus some others, though the complete family of homotopy spheres arising in this way is not known. Note that the diffeomorphism classes of homotopy spheres of dimension m which bound a parallelisable manifold of dimension $m+1$ forms a subgroup bP_{m+1} of the group of all diffeomorphism classes of homotopy m-spheres Θ_m. It is known that $bP_{odd} = 0$, bP_{4n+2} is either trivial or $\mathbb{Z}_2$ (with the non-trivial element where it occurs being a 'Kervaire sphere'), and that bP_{4n} is cyclic with order growing more than exponentially with n (see [KM]).

The Kervaire spheres are particularly simple to describe as the boundary of a plumbing: one simply plumbs together two copies of the tangent disc bundle of S^{2n+1} (see [LM; page 162]).

In order to describe a plumbing construction we can use a graph, with each bundle represented by a vertex and each plumbing by an edge joining the appropriate vertices. The case which is particularly important for the Carr argument is the plumbing of tangent disc bundles of S^{2n} according to the E_8 graph, (that is, the Dynkin diagram of the exceptional Lie group E_8). Call the resulting manifold W^{4n}. It is well known that the boundary Σ^{4n-1} of W is a homotopy sphere which generates the group bP_{4n}. Suppose the order of bP_{4n} is c_n. As this group is cyclic, it follows that the boundary connected sum of c_n of copies of W, $Y := \natural_{c_n} W$, will have boundary S^{4n-1}. Similarly the boundary connected sum of any number of copies of Y will also have boundary S^{4n-1}. In this way we can represent S^{4n-1} as the boundary of infinitely many different $4n$-manifolds. This observation is crucial for Carr's argument. (Clearly we can perform similar constructions for any homotopy sphere in bP_{4n}, a fact which Carr did not utilize. Moreover, according to [Wr1] we do not need to use the boundary connected sum to create this diversity of bounding manifolds: it turns out that plumbing alone will suffice.)

In general, the entire family of manifolds which arise as the boundaries of plumbings is unknown. However in the classical case of simply-connected plumbings of D^{2n}-bundles over S^{2n}, see [CW; Theorem C] for a complete answer.

Since we are primarily interested in the boundaries of plumbings, it is worth explicitly stating the boundary effect of the plumbing construction. Given disc bundles E_1 and E_2 as above, let us denote the plumbing of these bundles as $E_1 \Box E_2$. It is not difficult to see that

$$\partial(E_1 \Box E_2) = (\partial E_1 - D^m_M \times S^{n-1}) \cup (\partial E_2 - D^n_N \times S^{m-1}).$$

In the special case where the base manifolds M^m and N^n are spheres, it follows that the boundary effect of the plumbing is precisely equal to a surgery. A surgery of dimension p and codimension q on a manifold X^{p+q} involves removing the interior of the image of an embedding $\iota : S^p \times D^q \to X$ and replacing it with a copy of $D^{p+1} \times S^{q-1}$, which is glued in using the the map ι restricted to the boundary. The topological effects of surgery have been extensively studied, and it has proved an important technique for example in problems relating to the classification of manifolds (see [Ra] for instance). Moreover the behaviour of geometric phenomena such as positive scalar or positive Ricci curvature under surgery has played a crucial role in constructing examples and developing the theory of such curvature conditions. See for example [GL1], [St] and [Ro] for positive scalar curvature and [SY], [Wr1], [Wr2], [Wr3], [CW] for positive Ricci curvature. In particular note that by [GL1], positive scalar curvature metrics can be extended across any surgery of codimension at least 3.

§4.2.2 *Carr's argument*

There are two parts to Carr's argument: a topological part and a geometric part. We begin with the topological.

Fixing an orientation on the manifold Y constructed above, for $k, k' \in \mathbb{N}$ we form the oriented manifold $X_{k,k'} := (\natural_k Y) \cup_{S^{4n-1}} (-\natural_{k'} Y_{k'})$. It is clear that $X_{k,k'}$ is a closed manifold for any choice of k, k'.

We claim that for $k \neq k'$ the $\hat{A}$ genus of $X_{k,k'}$ is non-zero. The first step is to show that all the lower order Pontrjagin classes vanish (i.e. $p_i = 0$ for all $i < n$), and the second step is to show that the signature is non-zero. The third and final step is to compare the $\hat{A}$ and L-polynomials for manifolds of dimension $4n$. It is easily seen that if all the lower order Pontrjagin classes vanish, then $\hat{A}$ is proportional to the signature by some non-zero constant. The non-vanishing of the signature then guarantees the non vanishing of the $\hat{A}$-genus, as required.

To see the vanishing of the lower order Pontrjagin classes, first note that the manifold W (from which Y is constructed via boundary connected sums) has the homotopy type of a one-point union of eight copies of S^{2n} - the base spheres of the tangent disc bundles used in the plumbing of W. Using the Meyer-Vietoris sequence to handle the boundary connected sums, we see that the only possible $H^{4i}(X_{k,k'})$ for $i < n$ which could be non-zero is $H^{2n}(X_{k,k'})$. Therefore if n is even, i.e. $n = 2m$ for some m, we need to examine the Pontrjagin class $p_m(X_{k,k'})$. We claim that this is zero. To see this it suffices to evaluate this class on a basis for $H_m(X_{k,k'})$, namely the base spheres of each of the bundles used in the construction of $X_{k,k'}$. This evaluation is a local calculation, and thus (by naturality of the Pontrjagin classes under inclusion of the relevant disc bundle) depends only on $p_m(DTS^{2n})$ for each base sphere. But by the stability of the Pontrjagin classes under the addition of trivial bundles and the fact that TS^{2n} is stably trivial, we see that this (local) class, and therefore $p_m(X_{k,k'})$ must vanish as claimed.

To compute the signature, we note that the effect of the boundary connected sum on the intersection form is to split it into a direct sum of pieces, with each piece equal to the intersection form of W. The manifold W is well-known to have signature 8 (see [Br; V.2.8]). Thus the signature of Y is $8c_n$ and the signature of $\natural_k Y$ is $8kc_n$. Similarly the signature of $X_{k,k'}$ splits (via the Meyer-Vietoris sequence) into a sum of two signatures, though since we have reversed the orientation on $\natural_{k'} Y$ to form $X_{k,k'}$ we obtain $\sigma(X_{k,k'}) = 8c_n(k - k')$. Thus for $k \neq k'$ the signature and therefore the $\hat{A}$ genus is non-zero.

Now for the geometric part of Carr's argument. He claims that for any k, the manifold $\natural_k Y$ admits a positive scalar curvature metric $\bar{g}_k$ which is a product $dt^2 + g_k$ near the boundary, where g_k also has positive scalar curvature. To begin the argument we observe that we can put a submersion metric on any disc bundle which has positive scalar curvature and is a product near to the boundary. This follows from the O'Neill formulas for Riemannian submersions which show that the scalar curvature will be positive provided the scalar curvature of the fibres is sufficiently large (see [Be; 9.70]). Plumbing this disc bundle with a disc bundle over a sphere is then topologically equivalent to adding a handlebody to the original bundle. This should not be surprising since the boundary effect of such a plumbing is a surgery, as discussed in §4.2.1. Thus the sequence of plumbings required to construct the manifold W whose boundary generates bP_{4n} can be viewed as a sequence of handle additions to the tangent disc bundle of S^{2n}. Moreover the boundary connected sum construction, which Carr uses to produce the manifolds Y and $\natural_k Y$ from W, can also be viewed as the addition of a handlebody. Thus the geometric problem reduces to extending positive scalar curvature metrics across handlebodies in such a way that the resulting metric is still a product near the boundary. Carr gives an explicit though somewhat difficult metric construction which achieves this. Alternatively we could produce essentially the same result as follows. The plumbing of the manifold W provides a blueprint for constructing $S^{4n-1} = \partial W$ by a sequence of $(2n-1)$-dimensional surgeries starting with the tangent sphere bundle to S^{2n}. The boundary of $\natural_k Y$ can then be viewed as constructed from a number of copies of ∂W by the connected sum construction, that is, by a sequence of 0-surgeries. By [GL1] we can use this surgery description of S^{4n-1} coming from the construction of $\natural_k Y$ to equip $\partial(\natural_k Y)$ with a positive scalar curvature metric, which we will also label g_k. We now need to extend this boundary metric to a positive scalar curvature metric $\bar{g}_k$ on the whole of $\natural_k Y$, such that this metric is a product $dt^2 + g_k$ in a neighbourhood of the boundary. A method for doing this was given in [Ga], however this contained a mistake which was subsequently corrected and extended in [Wa1; Theorem 1.2].

Now assume we have positive scalar curvature metrics $\bar{g}_k$, $\bar{g}_{k'}$ on $\natural_k Y$ and $\natural_{k'} Y$, restricting to positive scalar curvature metrics g_k and $g_{k'}$ on the boundaries as above. The next step is to suppose that g_k and $g_{k'}$ belong to the same path component of positive scalar curvature metrics on $S^{4n-1} = \partial(\natural_k Y) = \partial(\natural_{k'} Y)$. This means we we can choose a path of positive scalar metrics h_t, $t \subset [0,1]$, so that the path is constant near the endpoints of $[0,1]$, and $h_0 = g_k$, $h_1 = g_{k'}$. A straightforward scalar curvature computation then shows that there is a smooth positive scalar curvature metric on $[0,a] \times S^{4n-1}$ for some $a > 0$ taking the form $f^2(t)dt^2 + h_t$, such that restricting to a neighbourhood of each boundary component gives $dt^2 + g_k$ at one end and $dt^2 + g_{k'}$ at the other. The function $f(t)$ is constant with value one near the endpoints of $[0,a]$, but very large in the middle.

It is now an elementary observation that if g_k and $g_{k'}$ belong to the same path component of $\mathcal{R}^+_{scal}(S^{4n-1})$, the space of positive scalar curvature metrics on S^{4n-1}, we can simply glue the metrics on $\natural_k Y$, $\natural_{k'} Y$ and $[0,a] \times S^{4n-1}$ along their isometric boundaries to obtain a smooth positive scalar curvature metric on the closed manifold $X_{k,k'}$. However $\hat{A}(X_{k,k'}) \neq 0$, and since the $\hat{A}$-genus obstructs positive scalar curvature in these dimensions we have a contradiction.

We conclude that the metrics g_k and $g_{k'}$ must belong to different path components of $\mathcal{R}^+_{scal}(S^{4n-1})$, and since there are infinitely many choices for k and k' we therefore have:

Theorem 4.2.2.1. *([Ca]) For each $n \geq 2$, $\mathcal{R}^+_{scal}(S^{4n-1})$ has infinitely many path components.*

In fact the above result can be generalized for any closed spin manifold X^{4n-1} with positive scalar curvature. We proceed as follows. For a fixed positive scalar curvature metric g_X on X, consider the product metric on $X \times [0,1]$, and form the boundary connected sum $X_k := X \times [0,1] \natural (\natural_k Y)$. Notice that the boundary of this manifold is $X \coprod (X \sharp S^{4n-1}) \cong X \coprod X$. As discussed above, with some local modification we can extend the product metric on $X \times [0,1]$ and the metric $\bar{g}_k$ on $\natural_k Y$ over the boundary connected sum to give a global positive scalar curvature metric which is a product near both boundary components. Denote the metric on the boundary component $X \sharp S^{4n-1}$ by $g_X \sharp g_k$. Now consider $X_{k'} := X \times [0,1] \natural (\natural_{k'} Y)$ for some k', equipped with the analogous metric. We can clearly glue X_k and $X_{k'}$ along each copy of $X \times \{0\}$ to obtain a new manifold with a smooth positive scalar curvature metric. As argued previously, if $g_X \sharp g_k$ and $g_X \sharp g_{k'}$ belong to the same path component of $\mathcal{R}^+_{scal}(X \sharp S^{4n-1})$, then we can join the two ends of $X_k \cup X_{k'}$ using a positive scalar curvature tube $(X \sharp S^{4n-1}) \times [0,a]$. This gives a smooth positive scalar curvature metric on the closed manifold $(X \times S^1) \sharp X_{k,k'}$ with $X_{k,k'} = (\natural_k Y) \cup_{S^{4n-1}} (-\natural_{k'} Y_{k'})$ as above. However

$$\begin{aligned}\hat{A}((X \times S^1) \sharp X_{k,k'}) &= \hat{A}(X \times S^1) + \hat{A}(X_{k,k'}) \\ &= \hat{A}(X_{k,k'})\end{aligned}$$

where we have used the additivity property of the $\hat{A}$-genus over connected sums together with the fact that the product metric $g_X + ds_1^2$ on $X \times S^1$ has positive scalar curvature, showing that $\hat{A}(X \times S^1) = 0$. Thus for $k \neq k'$ we have $\hat{A}((X \times S^1) \sharp X_{k,k'}) \neq 0$, contradicting the existence of a positive scalar curvature metric. We deduce:

Theorem 4.2.2.2. *For any closed spin manifold X^{4n-1} admitting a positive scalar curvature metric with $n \geq 2$, $\mathcal{R}^+_{scal}(X)$ has infinitely many path components.*

§4.3 *The work of Crowley-Schick*

We conclude this chapter by noting that the results of Hitchin described in §4.1 have recently been extended. In [CS], Crowley and Schick greatly improve Hitchin's results by uncovering exotic spheres 'deep' in the Gromoll filtration with non-zero α-invariant. Applying Hitchin's arguments to these spheres produces the following

Theorem 4.3.1. *([CS]) Let X^n be a closed spin manifold with positive scalar curvature metric. Then for all $j \in \mathbb{Z}$ with $8j+1 \geq n$ we have $\pi_{8j+1-n}\mathcal{R}^+_{scal}(X) \neq 0$. Specifically, this homotopy group can be shown to contain order 2 elements.*

The idea is roughly to form products involving an exotic 9-sphere with non-zero α-invariant with similar spheres in other dimensions. The resulting product can then be made into an exotic sphere by performing surgeries, which preserves the bordism type and hence the α-invariant. To show that such spheres live deep in the Gromoll filtration requires homotopy theoretic arguments using an alternative description of the Gromoll filtration given in terms of homotopy groups of the spaces $\mathrm{PL}(k)/\mathrm{O}(k)$, where $PL(k)$ is the group of piecewise linear homeomorphisms of $\mathbb{R}^k$ which fix the origin and $\mathrm{O}(k)$ is the usual orthogonal group.

Addendum

At the time of writing, new far-reaching results about the space of positive scalar curvature metrics have recently been announced by Botvinnik, Ebert and Randal-Williams [BER]. These results are essentially homotopy theoretic, but provide a unifying framework for, and indeed supersede many previous results about the space of positive scalar curvature metrics. This includes the results of Hitchin and Crowley-Schick discussed in this chapter.

5. The Kreck-Stolz s-invariant

In this chapter we introduce the s-invariant of Kreck and Stolz ([KS3]). For a Riemannian manifold (M, g) on which $s(M, g)$ is defined, this is an invariant of the path-component of the *moduli* space of positive scalar curvature metrics. With its introduction, it became possible for the first time to investigate the topology of moduli spaces of positive scalar curvature metrics (and in fact positive Ricci and non-negative sectional curvature metrics) on a wide variety of examples.

The s-invariant is not entirely straightforward to define. We will give the definition and explain the background in §5.2. The definition arises from the index theorem of Atiyah, Patodi and Singer ([APS1]), which we will discuss in §5.1 below (with more details provided in Appendix B). We will postpone applications of the s-invariant until the next chapter.

Note that the s-invariant is closely related to the Eells-Kuiper invariant [EK]. Indeed, suppose that for a Riemannian manifold (M^{4k-1}, g) the s-invariant is defined, then the Eells-Kuiper invariant is equal to $s(M; g)$ mod 1 in the case k is even and $s(M, g)/2$ mod 1 if k is odd. The Eells-Kuiper invariant is a smooth topological invariant, and thus reducing the s-invariant in this way removes the dependence on the metric.

§5.1 *The Atiyah-Patodi-Singer index theorem*

The Atiyah-Patodi-Singer index theorem generalizes the Atiyah-Singer index theorem discussed in §3.3 to manifolds with boundary. We will state the basic result for a *general* elliptic first order linear differential operator on a Riemannian manifold with boundary, specializing later to specific operators. The set-up for the general operator is as follows. Consider a manifold X^n, and complex vector bundles E and F over X. We will always suppose that these bundles are equipped with smoothly varying Hermitian inner products in the fibres. Consider a differential operator $D : \Gamma(E) \to \Gamma(F)$. Given local trivializations $E|_U \cong U \times \mathbb{C}^p$ and $F|_U \cong U \times \mathbb{C}^q$ over some neighbourhood $U \subset X$ equipped with a local coordinate system $(x_1, ..., x_n)$, we will assume D locally takes the following form:

$$D = \sum_{i=1}^{n} A_i(x) \frac{\partial}{\partial x_i}.$$

Here $x \in U$, and $A_i(x)$ is a $(q \times p)$-matrix of smooth complex-valued functions defined on U. Thus $A_i(x)$ determines a smoothly varying linear transformation of fibres $E_x \to F_x$.

In order to keep track of base spaces, bundles and boundary conditions, it will be convenient to follow the convention of [APS1] and write, for example, $C^\infty(X, E)$ instead of $\Gamma(E)$.

W. Tuschmann, D.J. Wraith, *Moduli Spaces of Riemannian Metrics*,
Oberwolfach Seminars 46, DOI 10.1007/978-3-0348-0948-1_5

Theorem 5.1.1. *([APS1; 3.10]) Let (X, g) be a compact Riemannian manifold with boundary Y, and let $D : C^\infty(X, E) \to C^\infty(X, F)$ be a linear first order elliptic differential operator. Suppose that the fibres of E and F are equipped with smoothly varying Hermitian inner products, which close to the boundary are independent of the normal parameter u. Restricted to a neighbourhood of Y, assume that D takes the form*

$$D = \sigma\Big(\frac{\partial}{\partial u} + A\Big)$$

where σ is a bundle isometry independent of u, and A is the obvious u-independent extension to the collar of an L^2 self-adjoint elliptic operator $A_0 : C^\infty(Y, E|_Y) \to C^\infty(Y, E|_Y)$. Let $C^\infty(X, E; P)$ denote the space of sections of E for which $P(f|_Y) = 0$, where

$$P : C^\infty(Y, E|_Y) \to C^\infty(Y, E|_Y)$$

denotes the projection onto the span of the non-negative eigenspaces of A_0. Then the restricted operator $D : C^\infty(X, E; P) \to C^\infty(X, F)$ has finite index given by

$$indD = \int_X \alpha_0(x)\, dx - \frac{h + \eta(0)}{2}.$$

Here $\alpha_0(x)$ is some function depending on the eigenvalues and eigenfunctions of the operators DD^* and D^*D, $h = dimkerA$ and $\eta(0)$ is the eta invariant of D.

Note that all dimensions here, i.e. the dimension h and the dimensions of kernel and cokernel in the index are *complex* dimensions.

We give an overview of the ideas behind the Atiyah-Patodi-Singer index theorem in Appendix B.

We need to say more about the function $\alpha_0(x)$ and the quantity $\eta(0)$ which appear in the index formula. We begin with $\eta(0)$, the so-called *eta invariant.*

First let us recall Fact 2 from §3.1 concerning self-adjoint elliptic operators D : $C^\infty(X, E) \to C^\infty(X, E)$: the eigenvalues of D are real, discrete and tend rapidly to infinity. The corresponding eigenspaces are finite dimensional and consist of smooth sections. Moreover $L^2(E)$ is a direct sum of the (mutually orthogonal) eigenspaces. We then define the *eta function* for such an operator D to be

$$\eta(z) := \sum_\lambda \operatorname{sign}(\lambda)|\lambda|^{-z}$$

where $z \in \mathbb{C}$ and the sum is over all the non-zero eigenvalues of D. Thus $\eta(z)$ is a complex valued function - where defined. A little complex analysis shows:

Lemma 5.1.2. *For all z with suitably large real part (say Re $z > R >> 0$) the above sum is convergent, and the resulting function is holomorphic on the half-space Re $z > R$.*

This holomorphic function $\eta(z)$ defined on a half-space can be extended in a unique manner across all of $\mathbb{C}$ to give a meromorphic function, also denoted $\eta(z)$. In future we will always assume that $\eta(z)$ refers to this extension.

Theorem 5.1.3. *([APS1; p60]) The eta function is holomorphic in a neighbourhood of $s = 0$, and moreover $\eta(0)$ is always real and finite.*

The eta invariant has an easy intuitive interpretation: it is roughly the difference between the number of positive and negative eigenvalues, i.e. it is a measure of *spectral asymmetry.*

A classic but simple example is to consider the (Dirac) operator $D := \frac{1}{i}d/d\theta$ acting on complex valued functions on the circle. The eigenfunctions then take the form $e^{in\theta}$ for $n \in \mathbb{Z}$, and so the eigenvalues are precisely all $n \in \mathbb{Z}$. Clearly, the spectrum is symmetric, and so $\eta(0) = 0$ in this case.

Eta invariants have been calculated explicitly in only a small number of cases, for example for some circle bundles and some homogeneous spaces. However we will not need to compute eta invariants explicitly for our purposes.

In §5.2 we will encounter eta invariants of different differential operators in the same formula. In order to keep track of these invariants we will use the notation $\eta(D)$ for the eta invariant $\eta(0)$ of a differential operator D.

We now turn our attention to α_0. The function α_0 is the constant term in the asymptotic expansion of $\sum e^{-t\mu}|\phi_\mu(x)|^2 - \sum e^{-t\mu'}|\phi'_{\mu'}(x)|^2$ where the μ and ϕ_μ are the eigenvalues respectively eigenfunctions of D^*D and μ' and $\phi'_{\mu'}$ are the eigenvalues respectively eigenfunctions of DD^* *on the double of* X, that is, on $X \cup_Y (-X)$.

In [Gi] Gilkey states (based on a result of Seeley [Se]) that $\alpha_0 = 0$ for odd dimensional X. In the light of this we can interpret η as a kind of index for odd dimensional manifolds.

Let us now specialize to those operators D which will be of most interest to us. Firstly, let (X^{2n}, g) be an even-dimensional, compact oriented spin manifold with Riemannian metric g, and let $\mathbb{S}$ be the irreducible complex spinor bundle over X. We consider again the Atiyah-Singer Dirac operator $D^+ : \Gamma(\mathbb{S}^+) \to \Gamma(\mathbb{S}^-)$ which we encountered in §3.3. This time, however, we must consider this operator in the case where X has non-empty boundary Y, and the Riemannian metric on X is a product in a collar neighbourhood of Y.

First note that near the boundary, this Dirac operator takes the required form

$$\sigma(\partial/\partial u + A).$$

Here σ is the map defined by Clifford multiplication by the unit inward normal vector at the boundary. In turn this means that $A = -e_{n+1} \cdot D_Y$ where D_Y is the Dirac operator on Y, in order for the Dirac operator in the collar to agree via restriction with that on Y.

For this Dirac operator, $\int_X \alpha_0\, dx$ is easily interpreted. Not surprisingly (if we compare with the Atiyah-Singer index theorem for closed manifolds in §3.3) we have $\int_X \alpha_0 = \int_X \hat{A}(p_1, ..., p_n)$ where the $p_i = p_i(X, g)$ are the Pontrjagin forms defined by the Levi-Civita connection of the metric g, and so in this case the index theorem reads

$$\mathrm{ind} D^+ = \int_X \hat{A}(p_1, ..., p_n) + \frac{h + \eta(0)}{2}.$$

The Atiyah-Patodi-Singer index for the Atiyah-Singer Dirac operator is sensitive to orientation. This is because changing the orientation switches the spinor bundles $\mathbb{S}^\pm$.

(Notice that the Clifford volume element changes sign, and in turn this switches the eigenspaces). Thus the kernel and cokernel of the Dirac operator are switched and so the index changes sign. The integral term here also switches sign, and therefore so must the $(h+\eta)/2$ term. Now h is always non-negative (though its value might vary when changing orientation: there is no reason why $\dim\ker D^+ = \dim\ker D^-$, since if these were equal the index would always be 0.). Thus η must change sign: if the boundary has positive scalar curvature (so $h=0$) then η becomes $-\eta$. However if h is non-zero then the changing of η must account for the h term.

The following computation will prove useful in applications of the Kreck-Stolz s-invariant. Consider oriented Riemannian spin manifolds (X,g) and (X',g') of dimension $4k$ with common boundary Y. Suppose that g and g' are product metrics near to the boundary, and both agree when restricted to Y. Assume further that the (common) metric on Y has positive scalar curvature so that $h=0$, and that the spin structures agree over Y. The index formula for the Dirac operator on the manifold $X \cup -X'$ gives

$$\begin{aligned}
\mathrm{ind}D^+ &= \int_{X\cup -X'} \hat{A}(p_i(X\cup -X', g\cup g')) \\
&= \int_X \hat{A}(p_i(X,g)) + \int_{-X'} \hat{A}(p_i(X',g')) \\
&= (\mathrm{ind}D_X^+ + \frac{1}{2}\eta_Y) + (\mathrm{ind}D_{-X'}^+ + \frac{1}{2}\eta_{-Y}) \\
&= (\mathrm{ind}D_X^+ + \mathrm{ind}D_{-X'}^+) + \frac{1}{2}(\eta Y - \eta Y) \\
&= \mathrm{ind}D_X^+ + \mathrm{ind}D_{-X'}^+
\end{aligned}$$

where we use the naturality of the Pontrjagin forms/classes on line 2, and the vanishing of h on line 3. Thus we obtain:

Lemma 5.1.4. *The index of $4k$-dimensional oriented Riemannian spin manifolds with boundary (with a product metric assumption in a neighbourhood of the boundary) is additive over isometric boundary components, provided the common boundary has positive scalar curvature.*

Finally, let us turn our attention to the signature operator discussed in §3.3, where we considered the Clifford bundle $Cl(X)$ for X^{4n} an oriented *closed* manifold (not necessarily spin), and defined the sub-bundles $Cl^{\pm}(X)$ to be the ± 1-eigenbundles under multiplication by $(-1)^n\omega$, where ω is the Clifford volume form. In this situation the Dirac operator splits as $D^{\pm} : \Gamma(Cl^{\pm}(X)) \to \Gamma(Cl^{\mp}(X))$ with D^+ being the signature operator, and we obtain the index theorem $\mathrm{ind}D^+ = \sigma(X)$, where $\sigma(X)$ is the signature of X.

Recall that for a closed manifold, the signature is defined to be the signature (that is, the number of positive eigenvalues minus the number of negative eigenvalues) of the symmetric form given by evaluating the cup product pairing for middle dimensional cohomology on the fundamental homology class. However, for a manifold X^{4n} with boundary Y, we define the signature in terms of the cup product *on the image of $H^{2n}(X,Y)$ in $H^{2n}(X)$* (under the inclusion map $(X,\emptyset) \hookrightarrow (X,Y)$).

It is convenient to view the signature operator as an operator on differential forms. In §2.1 we noted that we have a vector space isomorphism $Cl_n \cong \Lambda^*(\mathbb{R}^n)$, where $\Lambda^*(\mathbb{R}^n)$

is the exterior algebra on $\mathbb{R}^n$. More generally we have a bundle isomorphism $Cl(X) \cong \Lambda^*(X)$, where $\Lambda^*(X)$ is the bundle of exterior algebras associated to the tangent bundle, for which the sections are just the differential forms on X. Under this isomorphism the signature operator is just $d + d^* : \Omega^*_+(X) \to \Omega^*_-(X)$, where $\Omega^*_\pm(X)$ are the ± 1-eigenspaces of the involution $\tau : \Omega^*(X) \to \Omega^*(X)$ given on the space of p-forms $\Omega^p(X)$ by $\tau(\phi) = (-1)^{n+\frac{p(p-1)}{2}} * \phi$, where $*$ denotes the Hodge star operator. (We can view τ as an adjustment of the Hodge star operator so it squares to the identity.) It is easily checked that τ anti-commutes with $d + d^*$, so $d + d^*$ swaps the ± 1 eigenspaces.

On a manifold X with boundary Y, the signature operator takes the form $\sigma(\partial/\partial u + A)$ as required for the Atiyah-Patodi-Singer index theorem. Here $A\phi = (-1)^{n+p+1}(\epsilon * d - d*)\phi$ where ϕ is either a $2p$ form (in which case $\epsilon := 1$) or a $(2p-1)$-form (in which case $\epsilon := -1$). Let $B : \Omega^{even}(Y) \to \Omega^{even}(Y)$ be the restriction of A to the even dimensional forms on Y. We will also refer to B as a signature operator.

In the case where X^{4n} is a manifold with boundary, we have the following result ([APS1; 4.14]):

Theorem 5.1.5. *Let $(X^{4n}; g)$ be a compact oriented Riemannian manifold with boundary Y, and assume that in a collar neighbourhood of Y the metric is isometric to a product. Then*

$$\sigma(X) = \int_X L_n(p_1, ..., p_n) - \eta(B),$$

where $L_n(p_1, ..., p_n)$ is the nth L-polynomial in the Pontrjagin forms $p_i(X; g)$ (see §3.3).

§5.2 *The s-invariant*

In §5.1 we studied the Atiyah-Patodi-Singer index theorem for the index of a compact spin manifold W^{4k} with boundary M:

$$\text{ind}D^+(W, g_W) = \int_W \hat{A}(p_i(W, g_W)) - \frac{h(M) + \eta(D(M, g))}{2},$$

where the p_i are Pontrjagin forms and $h(M)$ the dimension of the space of harmonic spinors on M. For convenience we will refer to this theorem in future as 'APS'.

We consider positive scalar curvature metrics on manifolds of dimension $4k - 1$, and start by claiming that given such a metric g on a zero-bordant manifold M^{4k-1}, if we choose a bounding manifold W^{4k} and a corresponding extension of the metric g_W which is a product near the boundary M (but does not necessarily have positive scalar curvature everywhere), then $\text{ind}D^+(W, g_W)$ depends only on the connected component of g in $\mathcal{R}^+_{scal}(M)$. To see this, consider the effect on the index of smoothly deforming the metric. It is clear that the integral will vary smoothly, however the terms $h(M)$ and $\eta(D(M, g))$ might not. To see this latter point, consider the eigenvalues of the Dirac operator. These will clearly vary smoothly with the metric. However if a non-zero eigenvalue hits zero, or a zero eigenvalue moves away from zero, the value of $h(M)$ will jump by an integer amount. By a similar argument $\eta(D(M,g))$ might jump under the same circumstances. (Any jump in the eta invariant will also be by an integer amount, though this is not obvious from the definition of η.) There is also a hidden source of discontinuity in the APS formula. Recall that as part of the APS set-up we have to impose certain boundary conditions, and these also depend on the eigenvalues of the Dirac operator and in particular can vary discretely if an eigenvalues crosses zero. Taking all this into consideration, we see that if we can prevent any eigenvalues from hitting zero, the APS formula will vary smoothly under smooth metric deformation. A simple way in which to achieve this is to demand that all metrics during our deformation have positive scalar curvature at the boundary (as well as being a product in a neighbourhood of the boundary). It is a consequence of the Lichnerowicz theorem (see §3.4) that closed manifolds with positive scalar curvature have no harmonic spinors, i.e. the Dirac operator has no zero eigenvalues. Thus given two metrics g and g' belonging to the same path component of positive scalar curvature metrics on M, we can choose a smooth one parameter family of metrics on W extending a path of positive scalar curvature metrics from g to g' with the index varying smoothly over the path. But the index is an integer, and therefore must be constant, proving the claim.

Note also that by arguments in [APS2], if the metric on W also has positive scalar curvature the index vanishes (just as in the closed manifold case).

Thus the index of the Dirac operators on bounding manifolds offers the possibility of detecting non-triviality in $\pi_0\mathcal{R}^+_{scal}(M)$. There are problems with this approach however. Before we can even consider the potential difficulties with computing this index, we have the problem of finding bounding manifolds and suitable metric extensions. There is an underlying philosophical problem in that the information we desire about the manifold M is here presented in terms of some other manifold W. What we would like is to have an invariant which can be defined intrinsically in terms of (M,g). This turns out to be possible, but only with restrictions on the topology of M. In order to derive this invariant, we need to investigate $\mathrm{ind}D^+(W,g_W)$ in more detail.

Let us consider the $\hat{A}$-genus term in the above expression. This consists of a homogeneous polynomial of degree k in the Pontrjagin forms integrated over W. This polynomial consists of a non-zero scalar multiple of $p_k(W,g_W)$ plus sums of products of lower order Pontrjagin forms. With a view towards trying to express as much as possible of the index formula in terms of invariants of (M,g) alone, we make the following observation about the products of lower order terms in the polynomial $\hat{A}_k$. Under the assumption that all the real Pontrjagin classes of M vanish, by Stokes' Theorem the integral of these products of lower order terms over W can be expressed in terms of an integral over M, plus an element in the relative cohomology group $H^*(W,M;\mathbb{R})$. Specifically, we have:

Lemma 5.2.1. *([KS3; Lemma 2.7]) Let α and β be closed differential forms of positive degree on W whose restrictions to M are coboundaries, so there are forms $\hat{\alpha}, \hat{\beta}$ on M such that $d\hat{\alpha} = \alpha|_M$ and $d\hat{\beta} = \beta|_M$. Then*

$$\int_W \alpha \wedge \beta = \langle j^{-1}[\alpha] \cup j^{-1}[\beta], [W,M]\rangle + \int_M \hat{\alpha} \wedge \beta,$$

where $j^{-1}[\alpha], j^{-1}[\beta] \in H^*(W, M; \mathbb{R})$ are any preimages of the deRham cohomology classes $[\alpha], [\beta] \in H^*(W; \mathbb{R})$ under the map $j : H^*(W, M; \mathbb{R}) \to H^*(W; \mathbb{R})$ induced by inclusion, and $\langle \, , \, \rangle$ denotes the Kronecker product with the fundamental class of W. Moreover, the expression on the right-hand side is independent of all choices.

Kreck and Stolz introduce the notation $\int_M d^{-1}(\alpha \wedge \beta)$ to represent $\int_M \hat{\alpha} \wedge \beta$.

The significance of $\alpha|_M$ and $\beta|_M$ being coboundaries is that we want to get pre-images of α and β under the map $j : H^*(W, M; \mathbb{R}) \to H^*(W; \mathbb{R})$. The next map in the cohomology long exact sequence is $H^*(W; \mathbb{R}) \to H^*(\partial W; \mathbb{R})$ which is not in general going to be the zero map, but α and β map to zero because of the coboundary assumption and hence lie in the image of j. In terms of the index theorem, the coboundary assumption translates into the vanishing of the real Pontrjagin classes of the boundary, since the relevant differential forms represent real Pontrjagin classes of W.

Having established how to express products of lower order terms in the $\hat{A}$ polynomial in terms of quantities defined on the boundary, it remains to deal with the term involving p_k. The approach taken by Kreck and Stolz is to effectively eliminate it using the L-polynomial of the same order. This is also a polynomial in the Pontrjagin classes which contains a term involving a non-zero multiple of p_k, and thus the p_k terms coming from the two polynomials must be multiples. Let a_k denote the scaling factor so that $\hat{A} + a_k L$ has no p_k term. (It turns out that $a_k = 1/(2^{2k+1}(2^{2k-1} - 1))$.) In other words $\hat{A} + a_k L$ contains only products of lower order terms. On the other hand, the L-polynomial appears in the APS index formula applied to the signature operator:

$$\sigma(W) = \int_W L(p_i(W, g_W)) - \eta(B(M, g))$$

where B denotes the signature operator and $\eta(B(M, g))$ is its eta invariant. Combining all the above ideas gives the formula

$$\begin{aligned} \mathrm{ind} D^+(W, g_W) = \int_M d^{-1}(\hat{A} + a_k L)(p_i(M, g)) - \frac{h(M, g) + \eta(D(M, g))}{2} \\ - a_k \eta(B(M, g)) - t(W), \end{aligned}$$

where $t(W)$ is a topological term given by

$$t(W) = -\langle (\hat{A} + a_k L)(j^{-1} p_i(W)), [W, M] \rangle + a_k \sigma(W).$$

Notice that in the index expression, only the term $t(W)$ involves the manifold W: all other terms only depend on the boundary M. Thus in situations where $t(W)$ vanishes, the index is *completely determined by* (M, g). This suggests gathering together the terms which only depend on (M, g):

Definition 5.2.2. *(The Kreck-Stolz s-invariant.) Given a closed $(4k-1)$-dimensional spin manifold with vanishing real Pontrjagin classes and positive scalar curvature, let*

$$s(M, g) := -\frac{1}{2}\eta(D(M, g)) - a_k \eta(B(M, g)) + \int_M d^{-1}(\hat{A} + a_k L)(p_i(M, g)).$$

It is clear that s is invariant under spin structure preserving isometries of M. Moreover we make the following:

Claim 5.2.3. *The value of $s(M, g)$ only depends on the path component of g in $\mathcal{R}^+_{scal}(M)$.*

We will establish this claim later after we have considered certain other properties of the s-invariant.

A problem with the definition of s is that it is very hard to compute from the above formula (not least because eta invariants are difficult to compute). A more practical approach is given by the following

Observation 5.2.4. *Given (M, g) as in Definition 5.2.2, if M bounds a spin manifold W and g_W is any metric extending g (product near the boundary) then*

$$s(M, g) = indD^+(W, g_W) + t(W).$$

This is particularly useful in the case where an extension g_W exists with positive scalar curvature, for in this situation the index vanishes and we are left with $s(M, g) = t(W)$, with $t(W)$ computable from the topology of W. We will use this observation in applications. Notice that this also shows clearly that the s-invariant is always a rational number.

The existence of a spin manifold W bounding M is not guaranteed. In particular the vanishing of the real Pontrjagin classes is not sufficient to guarantee that M is a spin boundary. For example, consider the 'bad' exotic spheres with non-vanishing α-invariant (see §3.3). This means (since α is a ring homomorphism from the spin bordism ring) that these spheres cannot be spin boundaries. It is important to note, however, that the s-invariant is defined whether or not there exists a corresponding W.

Let us now return to the above claim that the value of $s(M, g)$ only depends on the path component of g in $\mathcal{R}^+_{scal}(M)$. To see this we need two facts: (i) the s-invariant is additive over disjoint unions, and (ii) the sign of the s-invariant is sensitive to orientation. Assuming these facts, we take $W = M \times I$ and let g' be any other metric in the same component of $\mathcal{R}^+_{scal}(M)$. We can put a positive scalar curvature metric on $M \times I$ which interpolates between g and g' (see §4.2.2). In this case we see that the index vanishes, and so we have $s(M \coprod (-M); g \coprod g') = t(M \times I) = 0$. (To see that $t(M \times I) = 0$ notice that the real Pontrjagin classes satisfy $p_i(W) = \sum_{j+k=i} p_j(M) \cup p_k(I) = p_i(M)$ as $p_k(I) = 0$ for all $k > 0$, and thus the real Pontrjagin classes of W are equal to the corresponding classes for M, which vanish by assumption. Moreover the signature of $M \times I$ also vanishes, as $M \times I \simeq M$ and so $H^{4k}(M \times I) = H^{4k}(M^{4k-1}) = 0$.) By facts (i) and (ii) above we can re-write $s(M \coprod (-M); g \coprod g')$ to obtain $s(M, g) - s(M, g') = 0$, establishing the claim.

It remains to justify facts (i) and (ii). To see additivity over disjoint unions we first observe that we can identify $H^n(M_1 \sqcup M_2)$ with $H^n(M_1) \oplus H^n(M_2)$ for all n via the isomorphism $\iota_1^* \oplus \iota_2^*$, where ι_1 and ι_2 are the inclusion maps into the disjoint union. We next observe that the tangent bundle $T(M_1 \sqcup M_2) = TM_1 \sqcup TM_2$, and from this and the naturality of Pontrjagin classes it follows that

$$p_i(M_1 \sqcup M_2) = p_i(M_1) \oplus p_i(M_2) \in H^{4i}(M_1) \oplus H^{4i}(M_2).$$

As $p_i(M_1) \cup p_j(M_2) = 0 \in H^{4(i+j)}(M_1) \oplus H^{4(i+j)}(M_2)$, we see that any polynomial in the Pontrjagin classes of a disjoint union splits into a sum of the individual polynomials for each connected component. Thus we have splitting of the integral term of s:

$$\begin{aligned}\int_{M_1 \sqcup M_2} & d^{-1}(\hat{A} + a_k L)(\{p_i(M_1 \sqcup M_2)\}) \\ &= \int_{M_1} d^{-1}(\hat{A} + a_k L)(\{p_i(M_1)\}) + \int_{M_2} d^{-1}(\hat{A} + a_k L)(\{p_i(M_2)\}).\end{aligned}$$

Turning our attention to the eta invariant of the Dirac operator, first observe that the spinor bundle over $M_1 \sqcup M_2$ splits as the disjoint union of the individual spinor bundles $SM_1 \sqcup SM_2$. The space of sections $\Gamma(S(M_1 \sqcup M_2))$ is easily seen to be $\Gamma(SM_1) \oplus \Gamma(SM_2)$. With respect to this splitting the Dirac operator takes the form

$$D = \begin{pmatrix} D_1 & 0 \\ 0 & D_2 \end{pmatrix}$$

where D_i is the Dirac operator on M_i. Thus the eigenvalues of D are the union of the eigenvalues of D_1 and D_2. The eta function $\eta(D)(z) = \sum_\lambda (\text{sign } \lambda)|\lambda|^{-z}$ where $z \in \mathbb{C}$ and λ runs over the set of eigenvalues for D, must then split as the sum $\eta(D_1)(z) + \eta(D_2)(z)$. Hence after analytically extending and evaluating at $z = 0$ we obtain a splitting of the eta invariant $\eta(D) = \eta(D_1) + \eta(D_2)$.

Finally, similar arguments apply to the eta invariant of the signature operator $B : \Omega^{even} \to \Omega^{even}$. As seen in §5.1, B has the following effect on a $2p$-form ϕ on a manifold of dimension $4k - 1$:

$$B(\phi) = (-1)^{k+p+1}(*d - d*)\phi.$$

Now $\Omega^{even}(M_1 \sqcup M_2) = \Omega^{even}(M_1) \oplus \Omega^{even}(M_2)$. It is clear that the exterior differential d respects this splitting, and it is easily checked that the Hodge $*$-operator does likewise. We conclude that the signature operator splits in similar fashion to the Dirac operator, and the splitting of the eta invariant for the signature operator then follows by the arguments above. Thus (i) is established.

To establish fact (ii), the sensitivity of s to orientation, we consider each of the terms on the right-hand side of the formula defining s. Firstly the integral term. The Pontrjagin forms of M or $-M$ are derived from the Levi-Civita connection of the metric, and this is independent of orientation. However the sign of integral does depend on the orientation of M, so changing the orientation of M results in the first term in the expression for s changing sign.

Dirac operators are independent of orientation. However the 'Dirac' operator under consideration here arises from restricting the Dirac operator on (W, g_W) to the boundary. As explained in §5.1, the Dirac operator on (W, g_W) takes the form $\sigma(\partial/\partial u + A)$ near the boundary, where σ is Clifford multiplication by the inward normal vector. At the boundary we have $A = -e_{4k} \cdot D_M$ where D_M is the Dirac operator on M, e_{4k} is the inward unit normal vector, and the dot indicates Clifford multiplication. Changing orientation can be achieved by replacing e_{4k} by $-e_{4k}$ in an ordered basis, and clearly this has the effect of changing A to $-A$. (See the comment at the bottom of page 60 of [APS].) As the boundary operator changes sign, so do the eigenvalues and consequently the eta invariant.

For the signature operator, note that d does not alter with the orientation but the Hodge star operator changes sign. Therefore B switches sign, and consequently so do its eigenvalues and eta invariant. Thus (ii) is established.

We remark that Gromov and Lawson [GL2] introduced a *relative* invariant $i(g, g')$ for positive scalar metrics g, g' on a closed $(4k-1)$-dimensional spin manifold M, defined as $i(g, g') = \mathrm{ind} D^+(M \times I, \bar{g})$ where $\bar{g}$ is any metric on $M \times I$ restricting to g, g' at the two boundary components (product near the boundary). By previous arguments, $i(g, g')$ is seen to only depend on the path components of g and g' in $\mathcal{R}^+_{scal}(M)$. It is also not difficult to see that if the real Pontrjagin classes of M vanish, then the s-invariant provides a corresponding absolute version of this, in the sense that $i(g, g') = s(M, g) - s(M, g')$.

Turning our attention now to moduli spaces, we need to consider the effect on the s-invariant if we pull back a metric on M via some diffeomorphism $M \to M$. The difficulty here is that such a diffeomorphism might not preserve the spin structure on M. In this case it is difficult to say what effect this would have on s. We can make things simpler if we only consider manifolds for which the spin structure is unique. For a given orientation, the corresponding spin structures are indexed by $H^1(M; \mathbb{Z}_2)$. Thus the spin structure is determined by the orientation if $H^1(M; \mathbb{Z}_2) = 0$. Even if this is the case, there is still an issue with the fact that the diffeomorphism might reverse the orientation, and thus change the sign of the s-invariant. We can overcome this issue by instead considering $|s|$. This leads to the following

Theorem 5.2.5. *([KS3; 2.14]) For a closed spin manifold M^{4k-1} with vanishing Pontrjagin classes and $H^1(M; \mathbb{Z}_2) = 0$, if the set of positive scalar curvature metrics on M is non-empty, then $|s(M; g)|$ is an invariant of the path component of $[g]$ in the moduli space of positive scalar curvature metrics.*

To conclude this section, we remark that the s-invariant behaves additively over connected sums in the following sense: if M_1 and M_2 admit positive scalar curvature metrics g_0 and g_1 then there is a an essentially canonical positive scalar curvature metric on $M_1 \sharp M_2$ due to Gromov and Lawson ([GL1]) which we will denote $g_1 \sharp g_2$.

Lemma 5.2.6. *Assuming the s-invariant is defined for both $(M_1; g_1)$ and $(M_2; g_2)$ we have*

$$s(M_1 \sharp M_2; g_1 \sharp g_2) = s(M_1; g_1) + s(M_2; g_2).$$

To see this we note that $M_1 \sharp M_2$ is bordant to $M_1 \sqcup M_2$ via $W := (M_1 \times I) \natural (M_2 \times I)$ where $\natural$ denotes the boundary connected sum. Thus W has oriented boundary components $M_1 \sharp M_2$ and $-M_1 \sqcup -M_2$. The additivity formula can then be obtained by computing $\mathrm{ind} D^+(W; \bar{g}) + t(W)$ where $\bar{g}$ is a positive scalar curvature metric extending $g_1 \sharp g_2$ (this exists by Gajer [Ga] and Walsh [Wa1]). As $\bar{g}$ has positive scalar curvature, the index term vanishes, so the computation reduces to understanding $t(W)$. Now $H^*(W) \cong H^*(M_1 \vee M_2) \cong H^*(M_1) \oplus H^*(M_2)$ for $* > 0$, where the first isomorphism follows from the fact that $W \simeq M_1 \vee M_2$, and the second from the Meyer-Vietoris sequence. By naturality of the Pontrjagin classes, the inclusion map $M_1 \sqcup M_2 \hookrightarrow W$ maps $p_i(W)$ to $p_i(M_1) + p_i(M_2)$. But this also induces the above isomorphisms in cohomology. Therefore the vanishing of the real Pontrjagin classes of M_1 and M_2 (by hypothesis) forces the vanishing of those for

W. As $t(W)$ can be completely described in terms of real Pontrjagin classes, we see that $t(W) = 0$. Thus the result is proved provided

$$\begin{aligned} s(\partial W, \bar{g}|_{\partial W}) &= s(M_1 \sharp M_2 \sqcup -M_1 \sqcup -M_2; g_1 \sharp g_2 \sqcup g_1 \sqcup g_2) \\ &= s(M_1 \sharp M_2; g_1 \sharp g_2) - s(M_1; g_1) - s(M_2; g_2), \end{aligned}$$

and this is true by facts (i) and (ii) above.

6. Applications of the s-invariant

In chapter 5 we saw that the s invariant $s(M,g) \in \mathbb{Q}$ is defined for any closed Riemannian manifold M of dimension $4k-1$ with vanishing real Pontrjagin classes and positive scalar curvature metric g. Moreover, if $H^1(M;\mathbb{Z}_2) = 0$ then $|s(M,g)|$ is an invariant of the path component of the moduli space of positive scalar curvature metrics containing g. Thus if we can find a collection of positive scalar curvature metrics on M which yield different s-invariants, we see immediately that the moduli space cannot be path-connected, and furthermore can hope to infer information about the extent of its disconnectedness. In general it is not too difficult to write down different positive scalar curvature metrics on a manifold which supports such metrics, however a more difficult challenge is to compute the corresponding s-invariants. The typical approach is topological: we look for a bounding manifold W, extend the metric to a metric g_W and exploit the formula

$$s(M,g) = \mathrm{ind}D^+(W, g_W) + t(W)$$

where the term $t(W)$ depends only on the topology of W. At first glance it might appear that we have just re-cast the problem of computing one difficult object, $s(M,g)$, in terms of another, $\mathrm{ind}D^+(W,g_W)$. However, with a careful choice of W it might be possible to find a metric extension g_W which has positive scalar curvature. In this case the index term vanishes, and so the s-invariant is equal to $t(W)$, which one hopes is computable via topological means. In this chapter we will see several instances where this strategy can be used effectively.

Although the immediate results one obtains from studying the s-invariant concern positive scalar curvature metrics, it is not difficult to use the s invariant to deduce path-connectedness results for moduli spaces of metrics satisfying stronger curvature conditions, such as positive Ricci curvature. For example if two Ricci positive metrics on a given manifold belong to different path components of the moduli space of positive scalar curvature metrics, it follows that they must also belong to different path components of the moduli space of positive Ricci curvature metrics. After discussing applications of the s-invariant to positive scalar curvature moduli spaces in §6.1, we turn our attention to positive Ricci curvature moduli spaces in §6.2. Finally, in §6.3 we look at applications of the s-invariant in the realm of sectional curvature.

§6.1 *Positive scalar curvature moduli spaces*

We begin this section by re-visiting the work of Carr described in §4.2. The main result of Carr discussed there was that the space of positive scalar curvature metrics on the sphere S^{4n-1} for $n \geq 2$ has infinitely many path components. It is now natural to ask whether the s-invariant can be used to deduce results about the moduli space of positive scalar curvature metrics on S^{4n-1}, and in particular to ask whether this moduli space also has infinitely many path components.

W. Tuschmann, D.J. Wraith, *Moduli Spaces of Riemannian Metrics*,
Oberwolfach Seminars 46, DOI 10.1007/978-3-0348-0948-1_6

Carr begins with a manifold-with-boundary W^{4n} with $\sigma(W) = 8$, which is constructed by plumbing eight copies of the tangent disc bundle of TS^{2n} according to the E_8-graph. The boundary of W is known to be an exotic sphere which generates the (cyclic) group of oriented diffeomorphism classes of homotopy spheres in dimension $4n-1$ which bound a parallelisable manifold. This is a finite group under the connected sum operation for which the order grows more than exponentially with n. Thus there is an integer c_n, growing rapidly with n, for which the boundary connected sum of c_n copies of W, $Y := \natural_{c_n} W$, has boundary $\partial Y = S^{4n-1}$ and signature $\sigma(Y) = 8c_n$. The manifold Y can be given a positive scalar curvature metric which is a product near the boundary. Denote this boundary metric on S^{4n-1} by g. We now use Y to define a closed manifold

$$X := Y \cup_{S^{4n-1}} D^{4n}.$$

The addition of a top-dimensional cell to Y clearly has no effect on the middle dimensional cohomology, and so we must have $\sigma(X) = \sigma(Y) = 8c_n$. Following exactly the same arguments as outlined in §4.2.2, the non-vanishing of the signature of X forces the $\hat{A}$-genus of X to be non-zero.

We now extend the metric on Y to a metric on X by extending over the disc D^{4n} in any way, and call the resulting metric $\bar{g}$. Note that $\bar{g}$ will not have positive scalar curvature globally over the disc.

We are now in a position to compute the s-invariant for (S^{4n-1}, g) via the formula

$$s(S^{4n-1}, g) = \mathrm{ind} D^+(D^{4n}, \bar{g}|_{D^{4n}}) + t(D^{4n}).$$

The contractibility of D^{4n} means that $t(D^{4n}) = 0$, and since the metric on Y has positive scalar curvature we also have $\mathrm{ind} D^+(Y, \bar{g}|_Y) = 0$. Therefore

$$s(S^{4n-1}, g) = \mathrm{ind} D^+(D^{4n}, \bar{g}|_{D^{4n}}) + \mathrm{ind} D^+(Y, \bar{g}|_Y).$$

We saw in §5.1 that the index is additive for manifolds with isometric positive scalar curvature boundaries, and hence

$$s(S^{4n-1}, g) = \mathrm{ind} D^+(X) = \hat{A}(X) \neq 0.$$

In §5.2 we also saw that the s-invariant is additive in the sense that $s(M \sharp M', g \sharp g') = s(M, g) + s(M', g')$, where $g \sharp g'$ is the 'canonical' positive scalar curvature metric arising from the Gromov-Lawson construction [GL1]. As a consequence we have $s(\sharp_k S^{4n-1}, \sharp_k g) = k\hat{A}(X)$ for any $k \in \mathbb{N}$. Since $\sharp_k S^{4n-1} \cong S^{4n-1}$ we deduce the following

Theorem 6.1.1. *The moduli space of positive scalar curvature metrics on S^{4n-1} $(n \geq 2)$ has infinitely many path components.*

Thus Carr's result *does* generalize to moduli spaces. In fact using similar arguments we can do much better:

Theorem 6.1.2. *Let M^{4n-1} $(n > 2)$ be any closed spin manifold with positive scalar curvature, vanishing real Pontrjagin classes and $H^1(M;\mathbb{Z}_2) = 0$. Then the moduli space $\mathcal{R}^+_{scal}(M)/\mathrm{Diff}(M)$ has infinitely many path components.*

Proof. Let h be any positive scalar curvature metric on M. Then by the additivity properties of s we have

$$s(M\sharp_k S^{4n-1}, h\sharp_k g) = s(M,h) + k\hat{A}(X),$$

where X is the manifold with $\hat{A}(X) \neq 0$ defined earlier in this section, and thus the metrics $h\sharp_k g$ on $M \cong M\sharp_k S^{4n-1}$ all yield different s-invariant values. □

This result should be compared with the theorem discussed at the end of §4.2.2, which asserted that for any spin manifold of dimension $4n-1$ $(n \geq 2)$ the space of positive scalar curvature metrics $\mathcal{R}^+_{scal}(M)$ has infinitely many path components. Thus using the s-invariant we are able to establish an analogous result about moduli spaces at the expense of having to restrict the topology of the manifolds under consideration.

§6.2 *Positive Ricci curvature moduli spaces*

At the start of this chapter we discussed how the s-invariant could be used to investigate the moduli spaces of metrics satisfying curvature conditions such as positive Ricci curvature, which are stronger than positive scalar curvature. In this section we will discuss several examples of positive Ricci curvature manifolds to which the approach can successfully be applied.

When considering a condition such as positive Ricci curvature in relation to positive scalar curvature there are many interesting questions one might ask. For example does *every* path component of (the moduli space of) positive scalar curvature metrics contain a component of positive Ricci curvature metrics? Can a given component of positive scalar curvature metrics contain more than one or even *infinitely* many components of positive Ricci curvature metrics? In the spirit of §6.1, we focus below on disconnectedness phenomena in the moduli space of positive Ricci curvature metrics.

§6.2.1 *The Einstein examples*

In [KS3], Kreck and Stolz use the s-invariant to produce the first examples of closed manifolds for which the moduli space of positive Ricci curvature metrics is highly disconnected, and the first examples for which the moduli space of positive sectional curvature metrics is not connected. Here we focus on the former, with a discussion of the positive sectional curvature examples postponed to §6.3.

For $k, l \in \mathbb{Z}$, let $M^7_{k,l}$ denote the principal S^1-bundle over $\mathbb{C}P^2 \times \mathbb{C}P^1$ with Euler class $lx + ky \in H^2(\mathbb{C}P^2 \times \mathbb{C}P^1)$, where the classes x and y correspond to generators of $H^2(\mathbb{C}P^2)$ respectively $H^2(\mathbb{C}P^1)$. Wang and Ziller [WZ] established the existence of Einstein metrics $g_{k,l}$ on the $M_{k,l}$ with positive Einstein constant (and thus also positive Ricci curvature).

These metrics are obtained as follows. Scale the standard product metric on $\mathbb{C}P^2 \times \mathbb{C}P^1$ by a factor μ_1 in the $\mathbb{C}P^2$ direction and by a factor μ_2 in the $\mathbb{C}P^1$ direction. Take the standard metric on S^1. By [Be; 9.59] given such base and fibre metrics, for each choice of principal connection there is a Riemannian submersion metric on the total space with totally geodesic fibres isometric to the given fibre metric. In our situation we choose the principal connection for which the curvature (with respect to the standard product metric on the base) is the harmonic form representing the cohomology class $lx+ky$. The resulting submersion metric can be shown to be Einstein with positive Ricci curvature for a suitable choice of the base scaling factors μ_1 and μ_2.

We will assume from now on that k and l are coprime. Using the homotopy long exact sequence of a fibration one can show that the $M_{k,l}$ are simply connected. In particular this means that $H^1(M_{k,l};\mathbb{Z}_2) = 0$. Moreover using the Gysin sequence in cohomology it can be shown (amongst other things) that $H^4(M_{k,l};\mathbb{R}) = 0$, (see [KS1; §4]), and thus all real Pontrjagin classes of $M_{k,l}$ vanish. So that the s-invariant is defined we must decide which $M_{k,l}$ are spin manifolds, and in order to use the formula $s(M,g) = \mathrm{ind}D^+(W,\bar{g})+t(W)$ we need to find for each spin $M_{k,l}$ a bounding spin manifold $W_{k,l}$ for which the spin structure extends that of $M_{k,l}$. Notice that there is an obvious candidate for such a $W_{k,l}$, namely the corresponding disc bundle.

Recall that a manifold is spin if and only if the first two Stiefel-Whitney classes vanish. For $M_{k,l}$ we have $w_1 = 0$ since $H^1(M_{k,l};\mathbb{Z}_2) = 0$. Instead of considering $w_2(M_{k,l})$, let us instead focus on $w_2(E_{k,l})$ where $E_{k,l}$ is the complex line bundle over $\mathbb{C}P^2 \times \mathbb{C}P^1$ corresponding to the circle bundle $M_{k,l}$. In other words E is the complex line bundle with Euler class $lx + ky \in H^2(\mathbb{C}P^2 \times \mathbb{C}P^1)$. The manifold $W_{k,l}$ bounding $M_{k,l}$ is then the unit disc bundle of $E_{k,l}$. Since $w_2(E_{k,l})$ is the modulo 2 reduction of the Euler class, we see that

$$w_2(E_{k,l}) = lx + ky \bmod 2 \in H^2(\mathbb{C}P^2 \times \mathbb{C}P^1;\mathbb{Z}_2) \cong \mathbb{Z}_2 \oplus \mathbb{Z}_2.$$

The tangent bundle of $E_{k,l}$ satisfies $TE_{k,l} \cong T(\mathbb{C}P^2 \times \mathbb{C}P^1) \oplus E_{k,l}$, and so $w_2(TE_{k,l}) = w_2(\mathbb{C}P^2 \times \mathbb{C}P^1) + w_2(E_{k,l})$. Now $\mathbb{C}P^2 \times \mathbb{C}P^1$ is not itself spin, since $w_2(\mathbb{C}P^2 \times \mathbb{C}P^1) = (1,0) \in \mathbb{Z}_2 \oplus \mathbb{Z}_2$, but this means that

$$w_2(TE_{k,l}) = (1,0) + (l \bmod 2, k \bmod 2).$$

Therefore $TE_{k,l}$ is spin if and only if l is odd and k is even. In this situation we see immediately by restriction that $W_{k,l}$ is spin, and this latter spin structure restricts to give a spin structure on $M_{k,l}$. Thus we obtain the desired spin structure on $M_{k,l}$ which can be extended over $W_{k,l}$, and therefore for a positive scalar curvature metric such as $g_{k,l}$ we see that the s-invariants $s(M_{k,l}, g_{k,l})$ are defined when l is odd and k even.

Theorem 6.2.1.1. *For l odd and k even,*

$$s(M_{k,l}, g_{k,l}) = -\frac{3k(l^2+3)(l^2-1)}{2^7 \cdot 7 \cdot l^2}.$$

It is easy to see that any metric $g_{k,l}$ can be extended to a positive scalar curvature metric $\bar{g}_{k,l}$ on $W_{k,l}$ which is a product near to the boundary. To do this simply replace the standard metric on S^1 by a rotationally symmetric metric on the disc D^2, and consider the corresponding submersion metric on $W_{k,l}$. According to the 'canonical variation' (see [Be; 9.70]), by uniformly shrinking the disc fibres (if necessary) we will produce a metric $\bar{g}_{k,l}$ with the desired properties. As a consequence we see that the index term in the formula $s(M_{k,l}, g_{k,l}) = \mathrm{ind}D^+(W_{k,l}, \bar{g}_{k,l}) + t(W_{k,l})$ vanishes, and thus the problem of computing the s-invariant in the above theorem reduces to finding the values of the topological terms $t(W_{k,l})$, which in turn can be found by direct calculation.

There is a diffeomorphism classification for the $M_{k,l}$ also due to Kreck and Stolz, see [KS1]. Comparing this classification with the above values of s, one can find infinitely many different (k,l) pairs for which the $M_{k,l}$ are diffeomorphic but the s-invariants are different. Thus we deduce

Theorem 6.2.1.2. *([KS3]) There are closed manifolds $M_{k,l}$ in dimension 7 which admit infinitely many Einstein metrics with positive Einstein constant with different s-invariant values. Consequently, for these manifolds $\mathcal{R}^+_{Ric}(M_{k,l})/\mathrm{Diff}(M_{k,l})$ has infinitely many path components.*

It is possible to say more about these examples, and we will return to the topic in §6.3.

§6.2.2 *Homotopy spheres*

As discussed in §4.2.1, homotopy spheres fall into two broad categories, those that bound and those that do not bound a parallelisable manifold. Let $bP_{n+1} \subset \Theta_n$ be the subgroup of oriented diffeomorphism classes of homotopy n-spheres which bound parallelisable manifolds of dimension $n+1$. Recall that this is a finite cyclic group for all n, and for $n \equiv 3 \bmod 4$ the order grows more than exponentially with n. As seen in §4.2.1 and §6.1, there is a manifold Y formed by plumbing copies of the tangent disc bundle DTS^{2n} for which the boundary generates bP_{4n}. As the group operation is connected sum, we see that any element of bP_{4n} can be constructed by a combination of plumbing disc bundles over S^{2n}, taking the boundary, and forming connected sums. In fact one can construct the same objects somewhat more cleanly:

Theorem 6.2.2.1. *([Wr1]) Every homotopy sphere in bP_{even} is the boundary of an explicit plumbing of disc bundles over S^{2n}.*

This was well-known in the case of bP_{4n+2} where there is at most one exotic sphere, and this can be constructed by simply plumbing two copies of DTS^{2n+1}. See for example [LM; page 162]. Thus the real content of the above result is in the case of bP_{4n}. (Note that $bP_{odd} = 0$.) The plumbing in this latter case involves copies of DTS^{2n} together with copies of another disc bundle over S^{2n} with Euler number 6. Although [Wr1] contains the existence result, the explicit details of the construction, including plumbing diagrams, are laid out in [Wr4]. Briefly, however, if Σ^{4n-1} represents a generator of bP_{4n}, then $k\Sigma = \Sigma\sharp...\sharp\Sigma$ arises as the boundary of a manifold constructed by plumbing the E_8 graph with all vertices representing DTS^{2n}, to $k-1$ copies of a certain graph involving both the disc bundles above. As a consequence of the finite order of bP_{4n} we deduce the following

Corollary 6.2.2.2. *Any given homotopy sphere in bP_{4n} can be expressed as the boundary of a manifold plumbed from disc bundles over S^{2n} in infinitely many different ways.*

We remark that very recently a powerful result was proved in [CW] (Theorem 2.4) showing that connected sums between the boundaries of large classes of plumbed manifolds can be represented explicitly as the boundary of a single plumbed manifold. In the case of the homotopy spheres above, this theorem gives alternative descriptions of these objects as boundaries of explicit manifolds plumbed from disc bundles over S^{2n}. In what follows, it does not matter which description is used.

Representing homotopy spheres as the boundary of plumbed manifolds is a crucial ingredient in the following theorem, bearing in mind that on the boundary, each plumbing is equivalent to a surgery. (See §4.2.1.)

Theorem 6.2.2.3. *([Wr1]) Every homotopy sphere which bounds a parallelisable manifold admits a metric of positive Ricci curvature.*

The idea behind the proof of this theorem is as follows. The plumbing diagrams discussed above provide a recipe for homotopy sphere construction by a sequence of surgeries. This surgery process for elements of bP_{4n} begins with the tangent sphere bundle STS^{2n}, with the first surgery being surgery on a fibre sphere. Surgery can be performed preserving $\mathrm{Ric} > 0$ under certain (reasonably tight) metric conditions: the normal disc bundle of the surgery sphere should be a product of a round metric on the sphere with a round disc for the normal fibre. Moreover the surgery sphere radius needs to be small compared to the normal discs. (Notice that this is the opposite of the situation which is most likely to occur naturally.) If we are performing surgery on a fibre sphere of a sphere bundle it is easy to arrange for these conditions to be met: we can choose a Ricci positive submersion metric with totally geodesic fibres which is a product in a neighbourhood of where the surgery is to be performed (see for example [Be; 9.59] in conjunction with [Be; 9.70]). We can then globally shrink the fibres (this can only increase the Ricci curvature) to achieve the required sphere to normal disc ratio, and thus we can extend the Ricci positive metric across the surgery. All the surgeries arising from the plumbings we are considering here can be handled (with some care) in a similar way. For more details about Ricci positive surgery see [Wr2] and also [SY].

Taken together, the above Corollary and Theorem indicate that on each homotopy sphere in bP_{4n} we can construct infinitely many different Ricci positive metrics, with one such metric arising from each of the surgery/plumbing descriptions of the sphere. A computation of s-invariants then yields:

Theorem 6.2.2.4. *([Wr4]) Both $\pi_0(\mathcal{R}^+_{Ric}(\Sigma)/Diff(\Sigma))$ and $\pi_0\mathcal{R}^+_{Ric}(\Sigma)$ are infinite when Σ^{4n-1} is any homotopy sphere in bP_{4n}.*

This result demonstrated for the first time that the moduli space of positive Ricci curvature metrics can be highly disconnected through an infinite range of dimensions. The Kreck-Stolz examples discussed above (as well as those in §6.3) are confined to dimension seven.

We have seen that calculating the s-invariant can present serious challenges. We will now outline the computation for these examples. We begin with a

Lemma 6.2.2.5. *If W is a parellelisable manifold with $\partial W = M$, which admits a positive scalar curvature metric $\bar{g}$ that in a neighbourhood of ∂W takes the form $dt^2 + g$, then*

$$s(M,g) = \frac{1}{2^{2n+1}(2^{2n-1}-1)}\sigma(W).$$

Notice in particular that the above Lemma applies to any homotopy sphere in bP_{4n}.

Proof (of the Lemma): Starting from the formula $s(M,g) = \mathrm{ind}D^+(W,\bar{g}) + t(W)$ we see that the index must vanish as $\bar{g}$ has positive scalar curvature, so it remains to find $t(W)$. Parallelisability means that all Pontrjagin classes vanish, and hence $t(W) = a_n\sigma(W)$, where a_n is known to be the coefficient in the Lemma. □

As in §6.1 and §4.2.2, we let W denote the manifold plumbed from copies of DTS^{2n} according to the E_8 graph, which has boundary $\Sigma \in bP_{4n}$. Let kW denote the plumbed manifold with boundary $k\Sigma \in bP_{4n}$ (which exists by [Wr1]). An easy algebraic topology argument shows that $\sigma(kW) = k\sigma(W) = 8k$. This means the above theorem on moduli spaces is proved provided we can extend the given Ricci positive metric on $k\Sigma$ to a positive scalar curvature metric on kW so that we have a product near the boundary. However the Ricci positive boundary metric here is quite different from the other positive scalar curvature metrics whose extensions we have discussed (for example in §4.2.2), essentially because the metric is created by a sequence of Ricci positive surgery constructions, and not via the Gromov-Lawson construction.

The general strategy is to take the construction surgery by surgery, considering the metric on each new piece arising from the surgery process explicitly, then extending the metric each time over the 'trace' of the surgery. It turns out that this can be done in a fairly natural manner, though the actual details are somewhat delicate. The next challenge is to arrange for the resulting metric to be a product near to the boundary. Unfortunately this does not happen automatically, and so a (positive scalar curvature) deformation is required. In turn, performing such a deformation is not a given, and requires a certain mean-curvature type condition at the boundary, but this condition can be shown to hold for all our examples. The full details are explained in [Wr4].

§6.2.3 *New results: highly connected manifolds*

A manifold M is said to be *highly connected* if it has dimension $2n$ or $2n+1$ and is $(n-1)$-connected, that is, if $\pi_0(M) = \pi_1(M) = \ldots = \pi_{n-1}(M) = 0$. Such manifolds were studied extensively from a topological point-of-view in the 1950s, 1960s and 1970s by Smale, Kervaire and Milnor, Wall and others (see for example [Sm], [KM], [W1], [W2], [Wi]). Despite their apparent topological simplicity, highly-connected manifolds constitute a rich class of manifolds. As an illustration of this fact note that in every dimension $n = 4k-1 \geq 7$, every finitely generated abelian group can arise as the integral cohomology group $H^{2k}(M)$ of such a manifold. (See [W2; Theorem 7] for $k \neq 2, 4$, and [Wi] for the remaining cases.) As the next theorem shows, in many cases highly connected manifolds admit Ricci positive metrics. For this theorem we recall that a manifold is said to be j-parallelisable if the tangent bundle restricted to some j-skeleton is trivial.

Theorem 6.2.3.1. *([CW]) Let $k \geq 2$, and let M^{4k-1} be a $(2k-2)$-connected manifold. If $k \equiv 1 \bmod 4$ assume further that M is $(2k-1)$-parallelisable. Then there is a homotopy sphere Σ^{4k-1} such that $M\sharp\Sigma$ admits a metric of positive Ricci curvature.*

With a little knowledge about the special cases of dimensions 7 and 11 one can deduce:

Corollary 6.2.3.2. *([CW]) All 2-connected 7-manifolds and all 4-connected 11-manifolds admit Ricci positive metrics.*

All the highly connected manifolds above which admit Ricci positive metrics can be constructed using plumbing, and the Ricci positive metrics can be created using Ricci positive surgery. Arguments about moduli spaces similar to those discussed in §6.2.2 for homotopy spheres can then by used to establish the following

Theorem 6.2.3.3. *Given M^{4k-1} as in the above theorem, there is a homotopy sphere Σ^{4k-1} such that $\mathcal{R}^+_{Ric}(M\sharp\Sigma)/\mathit{Diff}(M\sharp\Sigma)$ has infinitely many path components if (i) k is odd, or (ii) k is even and $H_{2k}(M;\mathbb{Z})$ consists entirely of torsion.*

Note that the extra restrictions here (compared to the previous theorem) arise so as to ensure that the s-invariant is defined. It is an open question as to whether the Ricci positive moduli space has infinitely many path components when k is even and H_{2k} is infinite.

§6.3 *Sectional curvature moduli spaces*

In [KS3], Kreck and Stolz give the first (and to date only) examples of closed manifolds with positive sectional curvature for which the moduli space of positive sectional curvature metrics can be shown to be disconnected. Specifically they exhibit two different positive sectional curvature metrics on a certain space for which the s-invariant takes different values.

The manifold in question is a 7-dimensional Aloff-Wallach space. The Aloff-Wallach manifolds $N^7_{k,l}$ are defined to be the quotient of SU(3) by a certain subgroup isomorphic to S^1. Let $i_{k,l} : S^1 \to \mathrm{SU}(3)$ denote the homomorphism

$$i_{k,l} : z \mapsto \begin{pmatrix} z^k & 0 & 0 \\ 0 & z^l & 0 \\ 0 & 0 & z^{-(k+l)} \end{pmatrix}.$$

Then $N_{k,l} := \mathrm{SU}(3)/i_{k,l}(S^1)$. These spaces form a topologically rich family of manifolds containing infinitely many homotopy types. In [KS2], Kreck and Stolz were able to give classifications up to both homeomorphism and diffeomorphism.

Let $g_{k,l}$ be the normal homogeneous metric on $N_{k,l}$, that is, the metric which makes the projection $\mathrm{SU}(3) \to N_{k,l}$ into a Riemannian submersion when SU(3) is equipped with a bi-invariant metric. Although this normal homogeneous metric has positive scalar curvature, it can be shown that the sectional curvature is only non-negative. In [AW], Aloff and

Wallach show that for certain k, l this metric can be deformed to one with strictly positive sectional curvature. The idea is as follows. For $z \in S^1$, the matrix

$$\begin{pmatrix} z^k & 0 \\ 0 & z^l \end{pmatrix}$$

is an element of U(2). There is an embedding of U(2) into SU(3) given by

$$A \mapsto \begin{pmatrix} A & 0 \\ 0 & (\det A)^{-1} \end{pmatrix}, \quad A \in \mathrm{U}(2).$$

Notice that $i_{k,l}(z)$ agrees with the image of $\mathrm{diag}(z^k, z^l)$ under this embedding. As a result we can view $N_{k,l} = \mathrm{SU}(3)/i_{k,l}(S^1)$ as the total space of the following fibre bundle:

$$\mathrm{U}(2)/i_{k,l}(S^1) \to \mathrm{SU}(3)/i_{k,l}(S^1) \to \mathrm{SU}(3)/\mathrm{U}(2).$$

(This is a lens-space bundle over $\mathrm{SU}(3)/\mathrm{U}(2)$.) The normal homogeneous metric $g_{k,l}$ clearly induces a Riemannian submersion structure on this bundle. (The invariance of $g_{k,l}$ under SU(3) restricts to give invariance under U(2).) It can be shown that for certain k and l, shrinking the fibres in this bundle creates strictly positive sectional curvature. As the shrinking process is a smooth operation, the resulting metric belongs to the same path component of positive scalar curvature metrics as $g_{k,l}$. The result is as follows:

Theorem 6.3.1. *If $kl(k+l) \neq 0$ there is a metric with positive sectional curvature on $N_{k,l}$ in the same path component of $\mathcal{R}^+_{scal}(N_{k,l})$ as $g_{k,l}$.*

In the light of this, we see that to compute the s-invariant of $N_{k,l}$ with this positive sectional curvature metric it suffices to compute $s(N_{k,l}, g_{k,l})$. Kreck and Stolz perform this calculation:

Theorem 6.3.2.

$$s(N_{k,l}, g_{k,l}) = \frac{kl(k+1)}{2^5 \cdot 7}.$$

The strategy is now the same as for the Wang-Ziller examples described in §6.2.1, namely to compare the above s-invariant values with the diffeomorphism classification of the Aloff-Wallach spaces. This yields the following

Theorem 6.3.3. *There are pairs (k, l) for which the corresponding Aloff-Wallach spaces are diffeomorphic but for which the s-invariants differ, so for these manifolds $N_{k,l}$ we have that $\mathcal{R}^+_{sec}(N_{k,l})/\mathit{Diff}(N_{k,l})$ is not path-connected.*

Kreck and Stolz only exhibit a single pair as the associated number theory is too difficult, though they mention the existence of two further examples.

A word should be said about the computation of the s-invariants $s(N_{k,l}, g_{k,l})$. This is very difficult and quite unlike the comparable computation for the examples in §6.2.1. The fundamental difference is that here there is no obvious choice of spin bounding manifold $W_{k,l}$ with which to work, as the natural spin structure on $N_{k,l}$ does not extend across the corresponding disc bundle. The strategy is roughly as follows. Suppose that W is *some* spin bounding manifold for an Aloff-Wallach space N, and let E denote the complex line bundle associated to N (viewing N as an S^1-bundle). Notice that the base of E is $B := \mathrm{SU}(3)/\mathrm{U}(2)$. The s-invariant for N with its normal homogeneous metric g is given by the usual formula $s(N,g) = \mathrm{ind}D^+(W, g_W) + t(W)$. The topological term t has some useful properties: it is additive when manifolds are glued along their boundaries, and for a closed manifold M it is easily checked that $t(M) = -\hat{A}(M)$. With these facts in mind we form the closed manifold $DE \cup_N W$, where DE is the disc bundle of E. The s-invariant formula can then be re-written as

$$s(N,g) = \mathrm{ind}D^+(W, g_W) + \hat{A}(DE \cup_N -W) + t(DE).$$

It turns out that the value of $\mathrm{ind}D^+(W, g_W) + \hat{A}(DE \cup_N -W)$ is independent of the choice of W, and moreover depends only on the bordism class of the pair (B, E) in a certain bordism group. Using some homotopy theory it can be shown that for given (B, E) there is a 'nice' bordant pair, consisting of a base manifold which is the total space of a complex projective bundle, together with the corresponding tautological complex line bundle. The s-invariant can then be computed in terms of this new pair, though this still requires a considerable calculation in algebraic topology.

Finally, we return to the examples $M_{k,l}$ of §6.2.1, and an enhancement of the Kreck-Stolz result which was pointed out in [KPT].

There is an alternative description of these manifolds as quotients $(S^5 \times S^3)/S^1$. Precisely, the manifold $M_{k,l}$ can be obtained as the quotient of $S^5 \times S^3$ by the S^1-action where $z \in S^1$ acts as

$$z \cdot (x, y) \mapsto (z^k \cdot x, z^l \cdot y).$$

The standard product metric on $S^5 \times S^3$ is invariant under this action, and so we obtain a metric on $M_{k,l}$ with non-negative sectional curvature and even positive Ricci curvature. Just as for the metrics in §6.2.1, it is easily seen that this metric on $M_{k,l}$ extends to a positive scalar curvature metric (product near the boundary) on the corresponding disc bundle $W_{k,l}$, and thus the computation of s-invariants again reduces to that of the topological terms $t(W_{k,l})$. We deduce

Theorem 6.3.4. *There are closed manifolds $M_{k,l}$ in dimension 7 for which the moduli space of non-negatively curved metrics $\mathcal{R}_{sec\geq 0}(M_{k,l})/\mathrm{Diff}(M_{k,l})$ has infinitely many path components.*

Addendum

At the time of writing, Dessai, Klaus and Tuschmann have just announced that there are infinite sequences of closed simply connected smooth manifolds with pairwise distinct homotopy type in dimensions $4k-1$ for all $k \geq 2$ for which the moduli space of non-negative sectional curvature metrics has infinitely many path components.

7. The Observer Moduli Space

In this chapter we introduce the observer moduli space. This is a variant of the usual moduli space of Riemannian metrics, and was first introduced by Akutagawa and Botvinnik in [AB]. The observer moduli space offers certain advantages over the original notion. In particular, information about the higher homotopy and homology groups of the observer moduli space is sometimes more easily available. In certain special situations we can use information about the topology of the observer moduli space to draw conclusions about the topology of the usual moduli space. We will introduce the basic ideas in §3.1 below, then discuss the key papers [BHSW] and [HSS] in the subsequent sections.

All manifolds under consideration in this chapter should be assumed compact, connected and without boundary, unless stated otherwise.

§7.1 *The basics*

To construct the moduli space of Riemannian metrics on a manifold M, we consider the action by pull-back of the diffeomorphism group $\mathrm{Diff}(M)$ on the space of metrics $\mathcal{R}(M)$, and form the quotient $\mathcal{R}(M)/\mathrm{Diff}(M)$. One 'difficulty' with this quotient operation is that the action of $\mathrm{Diff}(M)$ on $\mathcal{R}(M)$ is not free: for any $g \in \mathcal{R}(M)$, the isotropy group of the action is precisely the group of isometries of the metric g. The group of isometries is a Lie group, compact for M compact, typically of positive dimension.

In [AB], Akutagawa and Botvinnik had the idea of replacing the full group of diffeomorphisms with a certain subgroup which *does* act freely on $\mathcal{R}(M)$:

Definition 7.1.1. *Given a point $x_0 \in M$, the diffeomorphism group with observer, $\mathit{Diff}_{x_0}(M)$, is the subgroup of $\mathit{Diff}(M)$ consisting of diffeomorphisms $\phi : M \to M$ satisfying $\phi(x_0) = x_0$ and $D\phi|_{T_{x_0}M} = id_{T_{x_0}M}$.*

Lemma 7.1.2. *The group $\mathit{Diff}_{x_0}(M)$ acts freely on $\mathcal{R}(M)$.*

Proof. Suppose that $\phi \in \mathrm{Diff}_{x_0}(M)$ fixes an element $g \in \mathcal{R}(M)$, so ϕ is an isometry of (M, g). As ϕ fixes both the point x_0 and the tangent space at x_0, the fact that it is an isometry means that it must also fix all geodesics emanating from x_0. As any point of M belongs to a geodesic through x_0, we deduce that ϕ fixes all of M, i.e. $\phi = \mathrm{id}_M$. □

Definition 7.1.3. *The observer moduli space of M is the space $\mathcal{R}(M)/\mathit{Diff}_{x_0}(M)$. Similarly the observer moduli space of positive scalar curvature metrics on M is given by $\mathcal{R}^+_{scal}(M)/\mathit{Diff}_{x_0}(M)$.*

(For the record, an analogous idea appeared many years before [AB] in the paper [FG]. In this paper, Freed and Groisser are interested in the space $\mathcal{R}(M)$ for a compact oriented manifold M, and consider a subset $\mathcal{R}'(M)$ consisting of metrics which admit no non-trivial isometries. They then consider the quotient $\mathcal{R}'(M)/\mathrm{Diff}^+(M)$ under the group of orientation preserving diffeomorphisms of M. It is clear that the group $\mathrm{Diff}^+(M)$ acts freely on $\mathcal{R}'(M)$.)

W. Tuschmann, D.J. Wraith, *Moduli Spaces of Riemannian Metrics*,
Oberwolfach Seminars 46, DOI 10.1007/978-3-0348-0948-1_7

As a consequence of 7.1.1-7.1.3, we see that there is a principal $\mathrm{Diff}_{x_0}(M)$-bundle as follows:

$$\mathrm{Diff}_{x_0}(M) \to \mathcal{R}(M) \to \mathcal{R}(M)/\mathrm{Diff}_{x_0}(M).$$

Since $\mathcal{R}(M)$ is a contractible space, this shows that the observer moduli space is in fact a classifying space for principal $\mathrm{Diff}_{x_0}(M)$-bundles. We can therefore write:

$$\mathcal{R}(M)/\mathrm{Diff}_{x_0}(M) = \mathrm{BDiff}_{x_0}(M) \text{ and } \mathcal{R}(M) = \mathrm{EDiff}_{x_0}(M).$$

Thus up to bundle equivalence, principal $\mathrm{Diff}_{x_0}(M)$-bundles over a space X are in one-to-one correspondence with $[X, \mathrm{BDiff}_{x_0}(M)]$, the set of homotopy classes of maps $X \to \mathrm{BDiff}_{x_0}(M)$.

In the case of positive scalar curvature metrics, we obtain a similar fibration

$$\mathrm{Diff}_{x_0}(M) \to \mathcal{R}^+_{scal}(M) \to \mathcal{R}^+_{scal}(M)/\mathrm{Diff}_{x_0}(M).$$

Of course we cannot make any claims for the observer moduli space of positive scalar curvature metrics being a classifying space, since the space of positive scalar curvature metrics $\mathcal{R}^+_{scal}(M)$ is not in general contractible.

Note that we have long exact sequences of homotopy groups arising from the above fibrations. In particular, the fact that $\pi_k(\mathcal{R}(M)) = 0$ for all k means that

$$\pi_k(\mathcal{R}(M)/\mathrm{Diff}_{x_0}(M)) \cong \pi_{k-1}\mathrm{Diff}_{x_0}(M)$$

for all $k \geq 1$.

There is a third sequence which is of interest:

$$\mathcal{R}^+_{scal}(M) \to \mathcal{R}^+_{scal}(M)/\mathrm{Diff}_{x_0}(M) \to \mathcal{R}(M)/\mathrm{Diff}_{x_0}(M),$$

where the first map is the obvious quotient map, and the second is the inclusion map. Although this is not a fibration, it could be described as a 'homotopy fibration' on the grounds that it gives rise to a homotopy long exact sequence:

$$\begin{aligned} ... \to \pi_{k+1}(\mathcal{R}(M)/\mathrm{Diff}_{x_0}(M)) \to \pi_k(\mathcal{R}^+_{scal}(M)) \to \pi_k(\mathcal{R}^+_{scal}(M)/\mathrm{Diff}_{x_0}(M)) \\ \to \pi_k(\mathcal{R}(M)/\mathrm{Diff}_{x_0}(M)) \to \pi_{k-1}(\mathcal{R}^+_{scal}(M)) \to ... \end{aligned}$$

To see how this sequence arises, consider the commutative ladder formed from the homotopy long exact sequence of the two genuine fibrations above, with the obvious maps induced by inclusion providing the 'rungs' of the ladder:

$$\begin{array}{ccccccccc} ... & \to & \pi_k(\mathcal{R}^+_{scal}(M)) & \to & \pi_k(\mathcal{R}^+_{scal}(M)/\mathrm{Diff}_{x_0}(M)) & \to & \pi_{k-1}(\mathrm{Diff}_{x_0}(M)) & \to & ... \\ & & \downarrow & & \downarrow & & \| & & \\ ... & \to & 0 & \to & \pi_k(\mathcal{R}(M)/\mathrm{Diff}_{x_0}(M)) & \to & \pi_{k-1}(\mathrm{Diff}_{x_0}(M)) & \to & ... \end{array}$$

A little diagram chasing shows that we can replace the groups $\pi_{i-1}(\mathrm{Diff}_{x_0}(M))$ in the top sequence by the groups $\pi_i(\mathcal{R}(M)/\mathrm{Diff}_{x_0}(M))$ without losing exactness, thus obtaining the desired long exact sequence.

We now turn our attention to bundles with fibre M and structural group $\mathrm{Diff}_{x_0}(M)$.

As the total space $\mathrm{EDiff}_{x_0}(M)$ over $\mathrm{BDiff}_{x_0}(M)$ is equal to the space $\mathcal{R}(M)$ of all Riemannian metrics on M, a fibre of the universal bundle $\mathrm{EDiff}_{x_0}(M) \to \mathrm{BDiff}_{x_0}(M)$ can be viewed in two ways: it is non-canonically diffeomorphic to the Lie group $\mathrm{Diff}_{x_0}(M)$, but can also be thought of as a set of Riemannian metrics on M, any two of which differ (via pull-back) by an element of $\mathrm{Diff}_{x_0}(M)$. Associated to this bundle is the universal bundle with fibre M and structure group $\mathrm{Diff}_{x_0}(M)$:

$$M \to \mathcal{R}(M) \times_{\mathrm{Diff}_{x_0}(M)} M \to \mathcal{R}(M)/\mathrm{Diff}_{x_0}(M) = \mathrm{BDiff}_{x_0}(M).$$

Recall that the total space of this associated bundle, $\mathcal{R}(M) \times_{\mathrm{Diff}_{x_0}(M)} M$, is the quotient of $\mathcal{R}(M) \times M$ under the action of $\mathrm{Diff}_{x_0}(M)$ defined as follows: given any element $(h, x) \in \mathcal{R}(M) \times M$ and any $g \in \mathrm{Diff}_{x_0}(M)$ we have $g \cdot (h, x) = (hg^{-1}, gx)$. Now since h is a Riemannian metric on M and hg^{-1} is the pull-back of h via the map $g^{-1} : M \to M$, setting $\langle\, ,\, \rangle_y := h|_{T_y M}$ we have

$$hg^{-1}|_{T_{gx}M}(g_* v, g_* w) = \langle g_*^{-1} g_* v, g_*^{-1} g_* w\rangle_{g^{-1}gx} = \langle v, w\rangle_x$$

for any $v, w \in T_x M$.

We claim that the fibres of this universal M-bundle with structure group $\mathrm{Diff}_{x_0}(M)$ can be equipped with Riemannian metrics in a canonical way, giving what could be described as a 'universal metric'.

Fix a point $b \in \mathrm{BDiff}_{x_0}(M)$ and consider the fibre of $\mathrm{EDiff}_{x_0}(M) = \mathcal{R}(M)$ over b. We claim that for each metric h in this fibre and for each choice of $x \in M$, we obtain a unique inner product on the tangent space to the fibre of the associated universal M-bundle at the point $[h, x]$. To see this consider any vectors v', w' in this tangent space. These correspond uniquely under the quotient map to a pair of vectors v, w tangent to M at the point $(h, x) \in \mathcal{R}(M) \times M$, and similarly to vectors $g_* v, g_* w$ tangent to M at the point (hg^{-1}, gx) for any $g \in \mathrm{Diff}_{x_0}(M)$. But by the above, $(hg^{-1})_{gx}(g_* v, g_* w) = \langle v, w\rangle_x$, and so the inner product obtained by restricting h to $T_x M$ agrees with the inner product obtained by restricting hg^{-1} to the tangent space to M at gx. Thus the inner product $h_{T_x M}$ descends to give a well-defined inner product tangent to the fibre at $[h, x]$ in the associated M-bundle.

We can identify the associated bundle fibre over b with $\{[h, x] \,|\, x \in M\}$. (Consider h to be fixed here.) From the above we obtain a unique well-defined inner product on each tangent space to the fibre over b (which is independent of the choice of h). As $b \in M$ is arbitrary we can thus make the following

Observation 7.1.4. *The universal M-bundle with structure group $\mathrm{Diff}_{x_0}(M)$ comes with canonical Riemannian metrics in the fibres. Moreover any smooth map*

$$X \to BDiff_{x_0}(M)$$

defines simultaneously by pull-back both an M-bundle over X with structural group $\mathrm{Diff}_{x_0}(M)$, and a Riemannian metric on the fibres of this bundle.

Now the M-bundle over X with structure group $\mathrm{Diff}_{x_0}(M)$ is determined (up to equivalence) by the *homotopy class* of the pull-back map $X \to \mathrm{BDiff}_{x_0}(M)$. It is therefore clear that different maps in the same homotopy class will produce the same M-bundle, but with different fibre metrics. Conversely, given a family of metrics in the fibres of such a bundle, we identify each fibre non-canonically (i.e. up to action by an element of $\mathrm{Diff}_{x_0}(M)$) with a 'standard' copy of M, and pulling back the metric then gives a well-defined element in $\mathcal{R}(M)/\mathrm{Diff}_{x_0}(M) = \mathrm{BDiff}_{x_0}(M)$. Thus this family of metrics determines a map $X \to \mathrm{BDiff}_{x_0}(M)$. In turn this map can be used to re-construct both the the family of metrics and the bundle on which they are defined. This simultaneous creation of a bundle together with metrics on its fibres plays a crucial role in the work of Botvinnik, Hanke, Schick and Walsh ([BHSW]) which we will discuss next.

§7.2 *The work of Botvinnik, Hanke, Schick and Walsh*

The paper [BHSW] contains the first results about the homotopy groups of the observer moduli space, and shows that in many cases the higher homotopy groups are non-trivial. In very special situations it is possible to make deductions about the homotopy groups of the usual moduli space, but we will consider this later.

The main result concerning the observer moduli space is the following

Theorem 7.2.1. *Given $k \in \mathbb{N}$ there is an integer $N(k)$ such that for all odd numbers $n > N(k)$ and all manifolds M^n admitting a positive scalar curvature metric h, the group $\pi_i(\mathcal{R}^+_{scal}(M)/\mathit{Diff}_{x_0}(M), [h])$ is non-trivial when $i \leq 4k$, $i \equiv 0$ mod 4.*

The starting point for understanding this result is to consider the case $M = S^n$. Recall from §7.1 that $\mathcal{R}(M)/\mathrm{Diff}_{x_0}(M) = \mathrm{BDiff}_{x_0}(M)$, and so $\pi_i(\mathcal{R}(M)/\mathrm{Diff}_{x_0}(M)) = \pi_i(\mathrm{BDiff}_{x_0}(M))$. Now the groups $\pi_i(\mathrm{BDiff}_{x_0}(S^n))$ have been studied using algebraic K-theory, with the following being the relevant result for our purposes (see [FH]): let n be odd with $n > N(k)$ for some $k \in \mathbb{N}$ as above. Then for $i \leq 4k$ we have

$$\pi_i(\mathrm{BDiff}_{x_0}(S^n)) \otimes \mathbb{Q} = \begin{cases} \mathbb{Q} \text{ if } i \equiv 0 \bmod 4 \\ 0 \ \text{ otherwise.} \end{cases}$$

We need to use this non-triviality of rational homotopy groups for $\mathcal{R}(S^n)/\mathrm{Diff}_{x_0}(S^n)$ to deduce non-triviality for the homotopy groups of $\mathcal{R}^+_{scal}(S^n)/\mathrm{Diff}_{x_0}(S^n)$. This is achieved by the next result:

Theorem 7.2.2. *Given $k \in \mathbb{N}$, consider an odd integer $n > N(k)$, where $N(k)$ is as in 7.2.1. Then for $i \leq 4k$ we have that*

$$\iota_* \otimes \mathbb{Q} : \pi_i(\mathcal{R}^+_{scal}(S^n)/\mathit{Diff}_{x_0}(S^n)) \otimes \mathbb{Q} \to \pi_i(\mathcal{R}(S^n)/\mathit{Diff}_{x_0}(S^n)) \otimes \mathbb{Q}$$

is a surjection, where ι denotes the obvious inclusion of moduli spaces.

In the special case $M = S^n$ the statement of Theorem 7.2.1 now follows immediately from combining Theorem 7.2.2 with the non-triviality of $\pi_i(\mathrm{BDiff}_{x_0}(S^n)) \otimes \mathbb{Q}$. By the arguments at the end of §7.1 concerning the simultaneous creation of bundles and fibre metrics, it is not difficult to see when an element of $\pi_i(\mathcal{R}(S^n)/\mathrm{Diff}_{x_0}(S^n)) \otimes \mathbb{Q}$ is the image under $\iota_* \otimes \mathbb{Q}$ of an element in $\pi_i(\mathcal{R}^+_{scal}(S^n)/\mathrm{Diff}_{x_0}(S^n)) \otimes \mathbb{Q}$. Consider an element of the former group. This can be represented by a map $S^i \to \mathcal{R}(S^n)/\mathrm{Diff}_{x_0}(S^n)$, and such a map determines both an S^n-bundle over S^i as well as a collection of Riemannian metrics on the fibres. Saying that the lift to $\pi_i(\mathcal{R}^+_{scal}(S^n)/\mathrm{Diff}_{x_0}(S^n)) \otimes \mathbb{Q}$ exists then amounts to saying that this bundle admits a *fibrewise* positive scalar curvature metric, that is, a Riemannian metric on the vertical tangent bundle which restricts to a positive scalar curvature metric on each fibre. Thus the task is as follows: to show that there is a generating set for the group $\pi_i(\mathcal{R}(S^n)/\mathrm{Diff}_{x_0}(S^n)) \otimes \mathbb{Q}$ when $i \equiv 0 \bmod 4$, for which the S^n-bundle over S^i corresponding to each element in this generating set admits a fibrewise positive scalar curvature metric. Note that these bundles will be non-linear in general.

Finding fibrewise positive scalar curvature metrics is not straightforward. The standard approach to putting metrics on bundles, namely fixing a fibre metric, a base metric and a horizontal distribution in order to construct a Riemannian submersion (see [Be; 9.59]) will not work here. The problem is that we would need to find a metric on the fibre S^n which is invariant under the action of the structure group $\mathrm{Diff}_{x_0}(S^n)$. Now $\mathrm{Diff}_{x_0}(S^n)$ is non-compact, and it is not difficult to see that no invariant metric exists, thus a new approach is needed here.

It turns out that for $i \equiv 0 \bmod 4$, a generating set for $\pi_i(\mathcal{R}(S^n)/\mathrm{Diff}_{x_0}(S^n)) \otimes \mathbb{Q}$ can be chosen so that the resulting bundles are all 'Hatcher bundles'. This is significant since such objects have been shown to admit fibrewise Morse functions, that is, smooth functions on the total space which restrict to Morse functions on the fibres. A detailed description of such fibrewise Morse functions on Hatcher bundles is given in the work of Goette [Go]. Given such a function, one can attempt to construct positive scalar curvature metrics using a technique developed by Walsh in [Wa1].

The technique in question is a generalization of the classic Gromov-Lawson surgery procedure (see [GL1]), which for the sake of completeness we will briefly review. Gromov and Lawson show how to extend a positive scalar curvature metric across a surgery of codimension at least three. Consider a manifold W^{n+1} with boundary $M \sqcup M'$, that is, a bordism between the manifolds M and M'. We can decompose such a bordism into handlebodies, however such a decomposition is not unique. The possible decompositions correspond to Morse functions on W, that is, smooth functions for which the critical points are non-degenerate (i.e. the Hessian matrix is non-singular) and isolated, with the index (the number of negative eigenvalues of the Hessian matrix) determining the handle dimension. On the boundary each addition of a handle corresponds to a surgery, so given W as above and a positive scalar curvature metric on M, provided these surgeries have codimension at least three, the Gromov-Lawson procedure then produces a positive scalar curvature metric on M'. In fact, [Wa1; Theorem 2.5], we can produce a positive scalar curvature metric on all of W, which is a product near each boundary component. Thus given W and a positive scalar curvature metric on M, we obtain positive scalar curvature metrics on W and M' for each Morse function on W^{n+1} with critical points of index at most $n-2$. (This last requirement ensures that the associated surgery has codimension at least three.)

The work of Walsh allows this idea to be generalized as follows. Instead of a single bordism W, we can instead consider a bundle with fibre W. We assume that in a neighbourhood of the boundary components this bundle is trivial, so the boundary components are diffeomorphic to $B \times M$ and $B \times M'$ where B is the base of the bundle. Given a suitably well-behaved smooth function on the total space of the bundle which restricts to each fibre to give a Morse function with critical points of index at most $n-2$, together with a smooth family of positive scalar curvature metrics on M indexed by the points of B, we can construct a metric on the total space which restricts to each fibre to give a positive scalar curvature metric which is a product with respect to the normal parameter near each boundary component, and which agrees with the given family of metrics on the fibres of the $B \times M$ boundary component. (See [BHSW; Theorem 2.12].)

In order to apply these positive scalar curvature construction techniques to the problem in hand, we split each of the S^n-fibres of our Hatcher bundle into a union of hemispheres, which results in the Hatcher bundle decomposing as a union of two identical disc bundles. We initially consider just one of these disc bundles. By Goette, this bundle admits a fibrewise Morse function taking values in $[0, 1/2]$, with precisely three critical points in each fibre and value $1/2$ on the boundary. There is a unique minimum (with value 0) in each fibre, and it turns out that the bundle is trivial when restricted to a neighbourhood of these minima. Thus if we remove a neighbourhood of the minimum point from each fibre, we obtain a bundle over S^i with cylindrical fibre $S^{n-1} \times I$ for some interval I, with one of the two boundary components equal to $S^{n-1} \times S^i$. Equipping the S^{n-1} fibres of this boundary component with a round metric of some fixed radius, the above-mentioned technique of Walsh guarantees the existence of a metric on the cylinder-bundle with positive scalar curvature on the fibres, which is a product near the boundary components and agrees with the round metrics specified at one of the ends. The product structure of the metric near the other boundary component then allows us to metrically glue two copies of the cylinder-bundle along this boundary component, resulting in a fibrewise positive scalar curvature metric on the Hatcher bundle with two copies of $D^n \times S^i$ removed. As the fibre metrics in a neighbourhood of the two boundary components $S^{n-1} \times S^i$ are all round, we can then 'cap' the bundle, i.e. glue in two copies of $D^n \times S^i$, and extend the metrics in a standard 'torpedo' form over these caps so as to obtain fibrewise positive scalar curvature metrics on the Hatcher bundle itself. This is precisely what was required in order to establish Theorem 7.2.2 in the special case $M = S^n$.

The general case follows from the above case by taking fibre connected sums between the trivial bundle $S^i \times M \to S^i$ and the Hatcher bundle. From a metric viewpoint, we equip the fibres of the trivial bundle with the metric h. By using a small deformation if necessary, we can assume that the metric h takes a standard 'torpedo' form in a small disc. Since the Hatcher bundles contain an embedded trivial bundle $D^n \times S^i$ where the metric on the D^n fibres is also a torpedo, it is not difficult to see that the fibre connected sum can be constructed so as to have positive scalar curvature metrics on the fibres. It can be shown (using 'higher Franz-Reidermeister torsion') that the resulting bundle is non-trivial, and hence its classifying map $S^i \to \mathrm{BDiff}_{x_0}(M) = \mathcal{R}(M)/\mathrm{Diff}_{x_0}(M)$ represents a non-trivial element in $\pi_i(\mathcal{R}(M)/\mathrm{Diff}_{x_0}(M), [h])$. As we have constructed positive scalar curvature on the fibres of this bundle, the non-trivial element of $\pi_i(\mathcal{R}(M)/\mathrm{Diff}_{x_0}(M), [h])$ must lift to a non-trivial element in $\pi_i(\mathcal{R}^+_{scal}(M)/\mathrm{Diff}_{x_0}(M), [h])$, as required.

In some very special cases it is possible to deduce non-triviality of homotopy groups for the genuine moduli space of positive scalar curvature metrics. The result is as follows:

Theorem 7.2.3. *For any $k > 0$ there is a closed smooth manifold M admitting a metric h with positive scalar curvature such that $\pi_{4q}(\mathcal{R}^{+}_{scal}(M)/\mathit{Diff}(M), [h])$ is non-trivial for $0 < q \leq k$.*

This theorem follows by showing that the canonical map

$$\mathcal{R}^{+}_{scal}(M)/\mathrm{Diff}_{x_0}(M) \to \mathcal{R}(M)/\mathrm{Diff}(M)$$

can induce non-trivial homomorphisms on homotopy groups. The manifolds M to which this theorem applies are odd dimensional and non-spin. The construction of suitable manifolds M is a delicate undertaking, involving a considerable amount of algebraic topology.

§7.3 *The work of Hanke, Schick and Steimle*

The results in the paper [HSS] concern the higher homotopy groups of both the space of positive scalar curvature metrics as well as the observer moduli space. In many cases, these groups are shown to contain infinite order elements. The approach is broadly similar to that in [BHSW], in the sense that the results centre around the existence of bundles with very special properties, and the construction of fibre connected sums between these bundles and certain trivial bundles. The bundles are not the Hatcher bundles in [BHSW], rather, smooth bundles over spherical bases having $4n$-dimensional spin total spaces with non-zero $\hat{A}$-genus. Such bundles are of independent topological interest, since their existence demonstrates for the first time that the $\hat{A}$-genus is not multiplicative in bundles. It is well-known that for any product, $\hat{A}(M \times N) = \hat{A}(M)\hat{A}(N)$. However since $\hat{A}(S^k) = 0$ for all k, if the analogous statement for bundles were true, this would mean that any bundle over a sphere must have vanishing $\hat{A}$-genus, which is not the case. The result is as follows:

Theorem 7.3.1. *Given $k, l \geq 0$ there is an $N = N(k,l) \in \mathbb{N} \cup \{0\}$, such that for all $n \geq N$ there is a $4n$-dimensional smooth closed spin manifold P with $\hat{A}(P) \neq 0$, and a smooth fibre bundle*

$$F \hookrightarrow P \to S^k.$$

Moreover F can be chosen to be l-connected, and the bundle $P \to S^k$ can be assumed to have a smooth section $s : S^k \to P$ with trivial normal bundle.

The significance of the section with trivial normal bundle is that it allows fibre connected sums with trivial bundles. Locally, a neighbourhood of this section can be identified with the trivial disc bundle $S^k \times D^{4n-k} \to S^k$. Thus given a further trivial bundle $S^k \times N^{4n-k} \to S^k$ we can perform a fibre connected sum, that is, perform a connected sum between each copy of D^{4n-k} and the corresponding copy of N in a coherent way so as to produce a new smooth bundle. One immediate consequence of this is that in addition to the l-connectedness of the fibre F, we can further assume that the α-invariant of F vanishes. To see this we simply take $N = -F$ in the above, where $-F$ denotes a spin manifold representing the negative of $[F]$ in the spin bordism group. The significance of this observation is that such a manifold F must admit a positive scalar curvature metric by [St], assuming the dimension of F is at least 5 and $l \geq 1$. Notice that the non-vanishing of

$\hat{A}(P)$ prevents the total space of the bundle P from admitting a positive scalar curvature metric. This means that there can be no metric on P such that the restriction to each fibre is a positive scalar curvature metric, for if this was possible, we could equip P with a submersion metric using these fibre metrics, the round metric on the base S^k, and any horizontal distribution. By shrinking the fibres if necessary, the O'Neill formulas for Riemannian submersions show that the resulting metric would have positive scalar curvature, giving a contradiction.

The bundle P is used to construct a bundle of bordisms. By considering again the trivial sub-bundle $S^k \times D^{4n-k} \to S^k$ within P, we can embed a copy of two smaller discs D^{4n-k} joined by a 'strip' in each fibre:

$$S^k \times \left(D^{4n-k} \cup_{D^{4n-k-1}\times\{0\}} D^{4n-k-1} \times [0,1] \cup_{D^{4n-k-1}\times\{1\}} D^{4n-k}\right) \subset S^k \times D^{4n-k}.$$

Removing the interiors of both of these small $(4n-k)$-discs transforms each fibre F into a bordism between two copies of S^{4n-k-1}, and the bundle P into a bundle of bordisms with two boundary components each diffeomorphic to $S^k \times S^{4n-k-1}$. From this point we can construct a further - and final - bundle of bordisms by forming the fibre connected sum with the trivial bundle $S^k \times (M \times [0,1]) \to S^k$ by removing the 'strip' $D^{4n-k-1} \times [0,1]$ from the first bundle and a similar strip from the second along the $[0,1]$ component, then gluing the boundaries in the obvious way. Call the resulting bundle of bordisms E.

Hanke, Schick and Steimle then establish the following result about fibrewise positive scalar curvature metrics on bundles of bordisms, with a view to applying it to the bundle E just constructed. This result goes somewhat further than the corresponding theorem in [BHSW] in that only topological conditions are imposed, and not conditions on the critical points of fibrewise Morse functions. The set-up is as follows. Let $W \hookrightarrow Q \to B$ be a smooth bundle where ∂W has two connected components M_0 and M_1, and the structural group of Q is $\mathrm{Diff}(W; M_0, M_1)$ consisting of diffeomorphisms of W which restrict to the identity in a collar neighbourhood of the boundary components. Thus ∂Q consists of two boundary components $\partial_0 Q$ and $\partial_1 Q$, both of which have the structure of trivial bundles $B \times M_i \to B$. A collar neighbourhood of $\partial_i Q$ is therefore also a trivial bundle over B, with fibre $M_i \times I$ for some interval I.

Theorem 7.3.2. *If $\dim W \geq 2\dim B + 5$, the inclusion $M_1 \hookrightarrow W$ is 2-connected, and there exists a fibrewise positive scalar curvature metric on a collar neighbourhood of $\partial_0 Q \subset Q$ which is a fibrewise product, then this metric can be extended to a fibrewise metric of positive scalar curvature on Q, which is a fibrewise product on a collar neighbourhood of $\partial_1 Q$.*

The proof involves an analysis of fibrewise Morse functions on the total space of the bundle, as does the corresponding result in [BHSW]. However in this case more general types of critical points are considered creating a more general result.

Applied to the bundle E, Theorem 7.3.2 says that provided $4n - k \geq 2k + 5$, i.e. $4n \geq 3k + 5$, given an element in $\pi_k(\mathcal{R}^+_{scal}(M); g_0)$ for some fixed choice of positive scalar curvature metric g_0 on M, we can use a representative map $\phi : S^k \to \mathcal{R}^+_{scal}(M)$ to produce a fibrewise metric of positive scalar curvature on a collar neighbourhood of $\partial_0 E$ and then 'push' the metric through E to obtain a fibrewise positive scalar curvature metric on a collar neighbourhood of $\partial_1 E$. Correspondingly we obtain a new map $\phi' : S^k \to \mathcal{R}^+_{scal}(M)$. With a little extra work we can ensure that the image of this new map lies in the same path component of $\mathcal{R}^+_{scal}(M)$ as the original, and is based at the same metric g_0. The claim is then that the element $[\phi'] \in \pi_k(\mathcal{R}^+_{scal}(M); g_0)$ has infinite order, as a consequence of the special properties of the initial bundle P. To see this we consider a group homomorphism

$$\hat{A}_\Omega : \Omega_k^{spin}(\mathcal{R}^+_{scal}(M)) \to \mathbb{Z},$$

where $\Omega_k^{spin}(\mathcal{R}^+_{scal}(M))$ is the bordism group of k-dimensional spin manifolds equipped with smooth maps into $\mathcal{R}^+_{scal}(M)$, which is defined as follows. Consider the trivial bundle $S^k \times M \times [0,1] \to S^k$, and equip one boundary component with the product metric $ds_k^2 + g_0$ and the other end with the metric $ds_k^2 \oplus (g_x)_{x \in S^k}$ where $\{g_x \mid x \in S^k\}$ are the family of metrics on M determined by some map $S^k \to \mathcal{R}^+_{scal}(M)$. We extend these boundary metrics in any way over the interior of the trivial bundle to obtain a metric g, and set

$$\hat{A}_\Omega(\phi) = \operatorname{ind}(D_g),$$

where $\operatorname{ind} D_g$ is the Atiyah-Patodi-Singer index of the Dirac operator D_g on $S^k \times M \times [0,1]$. (This is equal to the Gromov-Lawson relative index $i(ds_k^2 + g_0, ds_k^2 \oplus (g_x)_{x \in S^k})$, see for example [LM; page 329].) It is not difficult to show that $\hat{A}_\Omega$ is a well-defined map on $\Omega_k^{spin}(\mathcal{R}^+_{scal}(M))$. Pre-composing this map with the canonical map $\pi_k(\mathcal{R}^+_{scal}(M); g_0) \to \Omega_k^{spin}(\mathcal{R}^+_{scal}(M))$ produces a homomorphism $\hat{A}_\pi : \pi_k(\mathcal{R}^+_{scal}(M); g_0) \to \mathbb{Z}$. Straightforward topological arguments can then be employed to show:

Proposition 7.3.3. *For the family of metrics ϕ' constructed above, we have*

$$\hat{A}_\pi(\phi') = \hat{A}(P) \neq 0.$$

As an immediate corollary of this we deduce:

Theorem 7.3.4. *Given $k \in \mathbb{N} \cup \{0\}$, there is a natural number $N(k)$ such that for each $n \geq N(k)$ and each spin manifold M^{4n-k-1} admitting a metric g_0 of positive scalar curvature, the homotopy group $\pi_k(\mathcal{R}^+_{scal}(M); g_0)$ contains elements of infinite order if $k \geq 1$, and infinitely many different elements if $k = 0$.*

There is a concept of 'geometric significance' for elements of $\pi_k(\mathcal{R}^+_{scal}(M); g_0)$ introduced in this paper, and defined as follows. An element $c \in \pi_k(\mathcal{R}^+_{scal}(M); g_0)$ is said to be *not* geometrically significant if the family of metrics on M it specifies can be realized by pulling-back a fixed metric on M via a family of oriented diffeomorphisms given by some continuous map $(S^k, *) \to (\mathrm{Diff}^+(M), \mathrm{id})$. Conversely, if c cannot be realized via such a pull-back, it is called geometrically significant. Note that the examples produced in [Hi] and [CS] are not geometrically significant by virtue of their construction. It is not difficult to show:

Lemma 7.3.5. *Given some $k \geq 1$, if M has the property that the total space of any oriented fibre bundle with fibre M and base S^{k+1} has vanishing $\hat{A}$-genus, then if an element $c \in \pi_k(\mathcal{R}^+_{scal}(M); g_0)$ is not geometrically significant, we have $\hat{A}_\pi(c) = 0$.*

The above Lemma can then be used to deduce an enhanced version of Theorem 7.3.4:

Theorem 7.3.6. *In the situation of Theorem 7.3.4, if in addition M satisfies the condition in Lemma 7.3.5, then we have that $\pi_k(\mathcal{R}^+_{scal}(M); g_0)$ contains infinite order elements for which all multiples are geometrically significant.*

So far in this section we have not mentioned the observer moduli space. In [HSS] this enters the picture by considering the map $\hat{A}_\pi$ in conjunction with the homotopy long exact sequence of the fibration

$$\mathcal{R}^+_{scal}(M) \hookrightarrow \mathcal{R}^+_{scal}(M)/\mathrm{Diff}_{x_0}(M) \to B\mathrm{Diff}_{x_0}(M).$$

If M satisfies the hypothesis of Lemma 7.3.5, then $\hat{A}_\pi$ can be factored through the *image* of the map

$$\pi_k(\mathcal{R}^+_{scal}(M); g_0) \to \pi_k(\mathcal{R}^+_{scal}(M)/\mathrm{Diff}_{x_0}(M), [g_0]),$$

induced by projection. To see this note that it is clear that such a factorization exists if and only if the kernel of the map $\pi_k(\mathcal{R}^+_{scal}(M); g_0) \to \pi_k(\mathcal{R}^+_{scal}(M)/\mathrm{Diff}_{x_0}(M); [g_0])$ is contained in the kernel of $\hat{A}_\pi$. Now elements in the former kernel are easily seen to be not geometrically significant, and thus by Lemma 7.3.5, such element must also belong to the kernel of $\hat{A}_\pi$. We deduce:

Theorem 7.3.7. *If M satisfies the hypothesis of Lemma 7.3.5, then the group*

$$\pi_k(\mathcal{R}^+_{scal}(M)/\mathit{Diff}_{x_0}(M); [g_0])$$

contains elements of infinite order.

In the case $k = 0$, exploiting the fact that the isotopy classes of spin-preserving diffeomorphisms form a finite-index subgroup of $\pi_0\mathrm{Diff}(M)$, we can deduce from Theorem 7.3.4 that the set $\pi_0(\mathcal{R}^+_{scal}(M)/\mathrm{Diff}(M))$ is infinite if the image of $\hat{A}_\pi$ is infinite in degree 0.

The above results can be further extended if we introduce the framed bordism group $\Omega^{fr}_k(\mathcal{R}^+_{scal}(M))$. The framed bordism group Ω^{fr}_k is the bordism group of k-dimensional manifolds which come equipped with an embedding into a sphere of suitably high dimension with trivial normal bundle, together with a choice of trivialisation for the normal bundle. A manifold represents an element in framed bordism if and only if its tangent bundle is stably parallelisable. (See [Ra; page 114].) In particular, the sphere S^k together with its canonical normal bundle trivialisation represent an element in Ω^{fr}_k. Moreover, given a map $\phi : S^k \to \mathcal{R}^+_{scal}(M)$, we obtain an element in the group $\Omega^{fr}_k(\mathcal{R}^+_{scal}(M))$. There is a canonical map

$$\pi_k(\mathcal{R}^+_{scal}(M); g_0) \to \Omega^{fr}_k(\mathcal{R}^+_{scal}(M)),$$

and also a canonical map between framed bordism and spin bordism

$$\Omega^{fr}_k(\mathcal{R}^+_{scal}(M)) \to \Omega^{spin}_k(\mathcal{R}^+_{scal}(M)).$$

The map $\hat{A}_\Omega$ can be pre-composed with this latter map to create a new homomorphism

$$\hat{A}_{fr} : \Omega^{fr}_k(\mathcal{R}^+_{scal}(M)) \to \mathbb{Z}.$$

It is straightforward to see that $\hat{A}_\pi$ is the composition of the first of the canonical maps above with $\hat{A}_{fr}$. It is therefore clear that in situations where the map $\hat{A}_\pi$ detects infinite order elements in $\pi_k(\mathcal{R}^+_{scal}(M); g_0)$, the images of these elements in $\Omega_k^{fr}(\mathcal{R}^+_{scal}(M))$ are similarly mapped to non-zero elements in $\mathbb{Z}$ by $\hat{A}_{fr}$.

The relevance of framed bordism to the current discussion is due to the fact that the Hurewicz map $\pi_k(\mathcal{R}^+_{scal}(M); g_0) \to H_k(\mathcal{R}^+_{scal}(M))$ can be factored through canonical maps as follows:

$$\pi_k(\mathcal{R}^+_{scal}(M); g_0) \to \Omega_k^{fr}(\mathcal{R}^+_{scal}(M)) \to \Omega_k^{spin}(\mathcal{R}^+_{scal}(M)) \to H_k(\mathcal{R}^+_{scal}(M)).$$

Now the map between the framed bordism group and the homology group is a rational isomorphism, and so the existence of infinite order elements in the framed bordism group means their images in $H_k(\mathcal{R}^+_{scal}(M))$ also have infinite order. Moreover, if M has the property that every oriented fibre bundle with fibre M and base B^{k+1} has $\hat{A}$-genus equal to the product $\hat{A}(M)\hat{A}(B)$, then the map $\hat{A}_{fr}$ factors through the image of the projection-induced map

$$\Omega_k^{fr}(\mathcal{R}^+_{scal}(M)) \to \Omega_k^{fr}(\mathcal{R}^+_{scal}(M)/\mathrm{Diff}_{x_0}(M)).$$

The above observations result in the following theorem:

Theorem 7.3.8. *The images of the infinite order elements in $\pi_k(\mathcal{R}^+_{scal}(M); g_0)$ from Theorem 7.3.4, and the geometrically significant infinite order elements from Theorem 7.3.6, all have infinite order in $H_k(\mathcal{R}^+_{scal}(M))$ under the Hurewicz map. Moreover, if M is simply connected and has the property that every oriented fibre bundle with fibre M and base B^{k+1} has $\hat{A}$-genus equal to the product $\hat{A}(M)\hat{A}(B)$, then the image of the map*

$$\pi_k(\mathcal{R}^+_{scal}(M); g_0) \to H_k(\mathcal{R}^+_{scal}(M)/\mathit{Diff}_{x_0}(M))$$

contains elements of infinite order.

8. A survey of other results

In this chapter we present a diverse selection of results about spaces and moduli spaces of metrics, which stand to some extent outside the themes developed in this book so far. The first section concerns the work of Botvinnik and Gilkey on moduli spaces of positive scalar curvature metrics for spin manifolds in odd dimensions with finite fundamental groups. This topic is the most closely related to subject matter presented earlier in that it involves index theory, and in particular utilizes the eta invariant in a significant way. The next topic is the work of Walsh on the space of positive scalar curvature metrics on spheres. This investigates structures on such spaces which are more subtle than those considered elsewhere in the current literature, and go beyond issues of connectedness and homotopy groups. The next two sections focus on low dimensions. The first of these looks at metrics of positive (Gaussian) curvature on $\mathbb{R}^2$, and in particular the work of Belegradek and Hu. The following section outlines the results of Codá Marques on the moduli space of positive scalar curvature metrics on three dimensional manifolds, and the space of such metrics on S^3. As a counterbalance to earlier sections, we conclude with a brief look at the work of Lohkamp on the (moduli) space of metrics with negative scalar and negative Ricci curvature.

§8.1 *The work of Botvinnik-Gilkey*

In this section we discuss the results of Botvinnik and Gilkey in the paper [BG]. The main result of this paper is as follows:

Theorem 8.1.1. *([BG]). Let M be a closed spin manifold of odd dimension ≥ 5 with finite fundamental group. Then if M admits a positive scalar curvature metric, the moduli space of such metrics $\mathcal{R}^+_{scal}(M)/\mathit{Diff}(M)$ has infinitely many path components.*

The major achievement here is to show that certain moduli spaces of positive scalar curvature metrics have infinitely many components in odd dimensions, where the index of the Dirac operator is either 0 or takes values in $\mathbb{Z}_2$. The different path components are detected by the eta invariant. This technique works by virtue of the fact that the eta invariant satisfies a certain additivity property and, crucially, takes values in $\mathbb{R}$.

Let $\Omega_n^{Spin,+}$ denote the spin bordism classes of closed n-manifolds with positive scalar curvature, for which the positive scalar curvature metrics are extendible across the bordism such that they are a product in a neighbourhood of the boundary. We will restrict attention to $n \geq 5$. Elements of this group take the form $[M, s, g]$ where s is a spin structure and g a positive scalar curvature metric.

Consider a closed spin manifold Y^{2k-1}, $k \geq 3$, which is the boundary of a compact manifold X^{2k} equipped with a compatible spin structure. Suppose further that X comes with a Riemannian metric g which is a product $dt^2 + g_Y$ near the boundary. Our starting point, just as for the Kreck-Stolz s-invariant, is the Atiyah-Patodi-Singer index formula for

W. Tuschmann, D.J. Wraith, *Moduli Spaces of Riemannian Metrics*,
Oberwolfach Seminars 46, DOI 10.1007/978-3-0348-0948-1_8

Riemannian manifolds with boundary (see [APS1], or §5.1 and Appendix B for details). Recall that in the case of the Dirac operator D^+ this formula is

$$\mathrm{ind} D^+(X;g) = \int_X \hat{A} - \frac{h(Y;g_Y) + \eta(Y;g_Y)}{2},$$

where $\hat{A}$ represents the $\hat{A}$-polynomial evaluated on the Pontrjagin forms of the metric g, $h(Y;g_Y)$ is the dimension of the space of harmonic spinors on the boundary, and $\eta(Y;g_Y)$ is the eta invariant on the boundary, which is a measure of the spectral asymmetry of the Dirac operator $D^+_{(Y;g_Y)}$. We would like to use the eta invariant to produce a group homomorphism $\Omega^{Spin,+}_{2k-1} \to \mathbb{R}$. As a pre-requisite, we should look to define a map on the set of closed compact positive scalar curvature $(2k-1)$-manifolds which vanishes on any representative of the zero element in $\Omega^{Spin,+}_{2k-1}$. Note that the eta invariant does not satisfy this requirement: from the formula above we see that if both X and Y have positive scalar curvature (so h and the index both vanish), we have

$$\eta(Y;g_Y) = 2\int_X \hat{A},$$

and this will be non-zero in general.

The key idea is to twist the spinor bundles by forming a tensor product with a flat bundle E (over X), and to consider the corresponding 'twisted' Dirac operator D^+_E. (Indeed the Dirac operator extends in a natural way to an operator on the tensor product with any Riemannian bundle over X. See [LM; page 138] for details of this construction.) For this operator we have an index formula (see [APS1; page 62])

$$\mathrm{ind} D^+_E(X;g) = \int_X \mathrm{ch}(E)\hat{A} - \frac{h_{E|_Y}(Y;g_Y) + \eta_{E|_Y}(Y;g_Y)}{2},$$

where $\mathrm{ch}(E)$ is the Chern character of E, and the 'E' subscripts indicate quantities which correspond to the twisting by E. Now by the twisted version of the Lichnerowicz theorem (which differs from the standard version by an added term in the curvature of E - see [LM; page 164]), the flatness of the bundle means that if X has positive scalar curvature (product near the boundary), the space of (twisted) harmonic spinors on both X and Y still vanish. This means that both $h_{E|_Y}(Y;g_Y)$ and $\mathrm{ind} D^+_E(X;g)$ vanish. Therefore assuming that X is equipped with a positive scalar metric (product near the boundary), the index formula reduces to

$$\eta_{E|_Y}(Y;g_Y) = 2\int_X \mathrm{ch}(E)\hat{A}.$$

Since E is flat all its rational Chern classes vanish, and so the Chern character $\mathrm{ch}(E) = \mathrm{dim} E$. We deduce that for any Riemannian manifold (Y, g_Y) representing the zero element in $\Omega^{Spin,+}_{2k-1}$, the quantity $\eta_{E|_Y}(Y;g_Y) - \dim(E|_Y)\eta(Y;g_Y)$ vanishes.

Now let Y^{2k-1} denote any closed spin manifold which admits a positive scalar curvature metric (so Y is not necessarily a spin boundary). Given two metrics on Y which belong to the same path component of positive scalar curvature metrics, it is well known (see for example §4.2.2) that the manifold $Y \times I$ can be equipped with a positive scalar curvature metric, a product near the boundary components, which interpolates between the two given metrics on Y. Fixing this metric, taking a flat bundle F over Y, and applying the index theorem to the manifold $X := Y \times I$ twisted with $F \times I$, it is straightforward to check that $\eta_F - \dim(F)\eta$ is in fact constant on each path component of positive scalar curvature metrics on Y. This follows from the fact that eta invariants are trivially additive over disjoint unions, and are sign-sensitive to orientation (see §5.2). Note that these ideas can be found in section 3 of [APS2].

More generally, given two flat bundles E_1 and E_2 over Y with equal dimension, the quantity $\eta_{E_1} - \eta_{E_2} = (\eta_{E_1} - \dim(E_1)\eta) - (\eta_{E_2} - \dim(E_2)\eta)$ is constant on each path component of positive scalar curvature metrics on Y, and moreover vanishes if (Y, g_Y) is zero bordant in $\Omega_{2k-1}^{Spin,+}$.

Let us consider flat bundles in more detail. A principal G-bundle over a manifold M is flat (that is, admits a principal connection with vanishing curvature) if and only if it is induced from the universal covering bundle of M (a principal $\pi_1(M)$-bundle) via a homomorphism $\sigma : \pi_1(M) \to G$. (Viewed as a principal $\pi_1(M)$-bundle, the universal cover is determined by a classifying map $M \to B\pi_1(M)$. Composing this with the map of classifying spaces $B\sigma : B\pi_1(M) \to BG$ determined by the homomorphism σ gives a map $M \to BG$. The corresponding principal G-bundle is precisely a flat principal G-bundle over M.) To construct flat $\mathbb{C}^m$ bundles over M (that is, complex vector bundles which admit a linear connection with vanishing curvature) we either choose $G = \mathrm{U}(m)$, or if G is some other fixed group it suffices to specify a representation $\rho : G \to \mathrm{U}(m)$. In the latter case the classifying map of a flat G-bundle $M \to BG$ can be composed with the induced map $B\rho : BG \to BU(m)$ to give a map $M \to BU(m)$, defining a principal $\mathrm{U}(m)$-bundle. From this we can produce a flat $\mathbb{C}^m$ vector bundle via the associated bundle construction. We will assume that the group G and the homomorphism $\sigma : \pi_1(M) \to G$ are fixed, and consider flat bundles as being specified by representations $\rho : G \to \mathrm{U}(m)$. In the sequel we will need to consider flat bundles over manifolds with boundary. Note that if $Y = \partial X$, then a flat bundle over X clearly restricts to give a flat bundle over Y, since a flat connection restricted to the boundary is still flat.

It is easy to see that the eta invariant $\eta(\rho)$, corresponding to the Dirac operator twisted by the flat bundle determined by ρ, is additive with respect to the direct sum of representations. Given such a sum $\rho_1 \oplus \rho_2$, the bundle E with which we twist the Dirac operator then splits as a sum $E = E_1 \oplus E_2$. In turn the Dirac operator splits as $D^+ = D_1^+ \oplus D_2^+$, and hence the set eigenvalues of D^+ is the union of the eigenvalues of D_1^+ and D_2^+ individually. It follows that $\eta(\rho_1 \oplus \rho_2) = \eta(\rho_1) + \eta(\rho_2)$. Thus for a fixed Riemannian spin manifold, η gives a homomorphism from the semigroup (under $\oplus$) of unitary representations $\mathrm{Rep}(G)$ to $\mathbb{R}$. If we formally complete $\mathrm{Rep}(G)$ to the group of representations $R(G)$, we can similarly extend η to a group homomorphism on $R(G)$ by setting $\eta(\rho_1 \oplus -\rho_2) := \eta(\rho_1) - \eta(\rho_2)$ for any virtual representation $\rho_1 - \rho_2$. Note that this makes sense from the point of view of the Atiyah-Patodi-Singer formula, since the Chern character can be viewed as a function with domain $K(M)$, that is, as a function acting on virtual bundles. In this context we also need to interpret the index and h as differences with respect to ρ_1 and ρ_2.

An alternative way of viewing the above constructions avoiding the language of virtual bundles and representations is to consider ordered pairs of representations, which correspond to a difference in the virtual case.

For $k \geq 3$, consider again a Riemannian spin manifold X^{2k} having boundary Y^{2k-1}, with the usual positive scalar curvature assumptions (that is, a positive scalar curvature metric on the bounding manifold which is a product near the boundary). Given an element $\rho \in R(G)$ of virtual dimension 0 (i.e. a pair of unitary representations having the same dimension), the arguments above concerning the quantity $\eta_{E_1} - \eta_{E_2}$ show that the twisted eta invariant for Y corresponding to ρ satisfies $\eta(Y)(\rho) = 0$. (Here η is really a function of Y, ρ, and both the spin structure and Riemannian metric on Y. However for convenience we will often suppress the full dependency from our notation.)

More generally, consider manifolds Y, Z in $\Omega_{2k-1}^{Spin,+}$. Suppose these are bordant via a manifold X, so that $\partial X = Y \coprod -Z$. It is easy to see that the index formula can be generalized to the case where there is more than one boundary component. In this case we have

$$\mathrm{ind} D^+ = \int_X \hat{A} - (h(Y) + \eta(Y))/2 - (h(Z) + \eta(-Z))/2$$

in the non-twisted case. Moreover if we fix a homomorphism $\sigma : \pi_1(X) \to G$ and a representation $\rho : G \to U(m)$, the analogous index formula continues to hold in the resulting twisted case. (As previously noted, the flat vector bundle over X determined by σ and ρ restricts to give flat bundles over the boundary components.) Following the rationale above, given $\rho \in R(G)$ of virtual dimension 0, we see that under the usual positive scalar curvature assumptions we obtain $\eta(Y)(\rho) = \eta(Z)(\rho)$.

With this in mind we consider a more delicate bordism framework where we also take into account maps from the manifolds to BG, together with the obvious notion of bordism for such maps. This leads to the bordism groups $\Omega_*^{Spin}(BG)$ and $\Omega_*^{Spin,+}(BG)$. If G is a discrete group (and in particular if G is finite), then there is a bijection between the set of homotopy classes $[M, BG]$ and $\mathrm{Hom}(\pi_1(M), G)$ given by the induced map on fundamental groups. Consequently, for G discrete we can view $\Omega_*^{Spin}(BG)$ as spin bordism which takes into account homomorphisms from fundamental groups to G, or equivalently as spin bordism incorporating flat G-bundle structures over the manifolds. (Explicitly, this bordism notion is as follows. Let s and s' be spin structures on M^n respectively N^n, and let $\sigma : \pi_1(M) \to G$ and $\sigma' : \pi_1(N) \to G$ be homomorphisms. Then (M, s, σ) is bordant to (N, s', σ') if and only if there is a spin manifold $(W^{n+1}, \tilde{s})$ providing a spin bordism between (M, s) and (N, s'), and a homomorphism $\tilde{\sigma} : \pi_1(W) \to G$ such that $\sigma = \tilde{\sigma} \circ (\iota_M)_*$ and $\sigma' = \tilde{\sigma} \circ (\iota_N)_*$, where ι_M and ι_N denote the inclusion maps of M respectively N into W.)

From now on we will assume that G is a finite group. Our considerations above show that we have a well-defined function

$$\eta : \Omega_{2k-1}^{Spin,+}(BG) \otimes R_0(G) \to \mathbb{R},$$

where $R_0(G)$ is the subgroup of $R(G)$ with virtual dimension 0.

Now suppose Y and Z represent different bordism classes in $\Omega_{2k-1}^{Spin,+}(BG)$. Notice that $Y\sharp Z$, which has positive scalar curvature by Gromov-Lawson, is bordant to $Y\coprod Z$. Thus by applying our twisted virtual index theorem to this bordism we have $\eta(Y\sharp Z)(\rho) = \eta(Y)(\rho) + \eta(Z)(\rho)$ for any $\rho \in R_0(G)$. As the group operation in bordism can be taken to be the connected sum, we see that for fixed $\rho \in R_0(G)$, η gives in fact an additive homomorphism $\eta(\rho) : \Omega_{2k-1}^{Spin,+}(BG) \to \mathbb{R}$.

Suppose that g and g' are two positive scalar curvature metrics on M^{2k-1} which belong to the same connected component of $\mathcal{R}_{scal}^{+}(M)$. Now $M\times I$ can be equipped with a positive scalar curvature metric which restricts to dt^2+g and dt^2+g' near the boundary components, and since $\eta(\rho)$ is an invariant of $\Omega_{2k-1}^{Spin,+}(BG)$ we see that $\eta(M,g)(\rho) = \eta(M,g')(\rho)$. Thus $\eta(\rho)$ is in fact constant on each path component of positive scalar curvature metrics on M.

Since G is finite it is known that $\Omega_{2k-1}^{Spin}(BG)$ is a finite group. Thus taking $[M,s,\sigma]$, $[N,s',\sigma']$ in $\Omega_{2k-1}^{Spin}(BG)$, there is $n \in \mathbb{N}$ such that $n[N,s',\sigma'] = 0 \in \Omega_{2k-1}^{Spin}(BG)$, and hence $[M,s,\sigma] + rn[N,s',\sigma'] = [M,s,\sigma]$ for all $r \in \mathbb{Z}$. For clarity here, let us consider the class $[M,s,\sigma] + rn[N,s',\sigma']$. This is represented by the manifold $M\sharp_{i=1}^{rn} N$ equipped with the obvious spin structure s'' formed from s and s', so in particular we have that $(M\sharp_{i=1}^{rn} N, s'')$ is spin bordant to (M,s). The fundamental group of $M\sharp_{i=1}^{rn} N$ is simply the free product $\pi_1(M) * \pi_1(N) * \cdots * \pi_1(N)$. Notice that given maps σ and σ' as above, there is an induced homomorphism from this free product to G. Call this $\sigma'' : \pi_1(M\sharp_{i=1}^{rn} N) \to G$. The statement $[M,s,\sigma]+rn[N,s',\theta] = [M,s,\sigma]$ can then be re-expressed as $[M\sharp_{i=1}^{rn} N, s'', \sigma''] = [M,s,\sigma]$.

Given a bordism $[M,s,\sigma] = [\bar{M},\bar{s},\sigma]$ in $\Omega_{2k-1}^{Spin}(BG)$, suppose that $\bar{M}$ admits a positive scalar curvature metric $\bar{g}$, M is connected and $\sigma : \pi_1(M) \to G$ *is an isomorphism*. By work of Miyazaki [Mi] and Rosenberg [Ro], it is possible to 'push' the positive scalar curvature metric $\bar{g}$ through the bordism to get a positive scalar curvature metric g on M, such that $[M,g,s,\sigma] = [M,\bar{g},\bar{s},\bar{\sigma}]$ in $\Omega_{2k-1}^{Spin,+}(BG)$.

Now suppose that the manifold N above has a positive scalar curvature metric g' and a corresponding non-zero $\eta(\rho)$ for some $\rho \in R_0(G)$. If M, as in the paragraph above, has a positive scalar curvature metric g, then for each $r \in \mathbb{Z}$ we have a positive scalar curvature metric on $\bar{M} := M\sharp_{i=1}^{rn} N$ by Gromov-Lawson, and by Miyazaki and Rosenberg we obtain a family of positive scalar curvature metrics g_r on M. By the additivity of $\eta(\rho)$ we have $\eta(M,g_r,s,\sigma)(\rho) = \eta(M,g,s,\sigma)(\rho) + rn\eta(N,g',s',\sigma')(\rho)$. Clearly the values of $\eta(M,g_r,s,\sigma)(\rho)$ will be different for different r if $\eta(N,g',s',\sigma')(\rho) \neq 0$. As $\eta(\rho)$ is constant on each path component of positive scalar curvature metrics on M, we can then deduce that $\mathcal{R}_{scal}^{+}(M)$ must have infinitely many path components.

Assuming that $\pi_1(M)$ is non-trivial and *finite*, let $H \subset \mathrm{Diff}(M)$ be the subgroup of orientation and spin structure preserving diffeomorphisms which induce the identity map on $\pi_1(M)$. Notice that for $h \in H$, $\eta(M,h^*g_r,s,\sigma)(\rho) = \eta(M,g_r,s,\sigma)(\rho)$, so if $\mathcal{R}_{scal}^{+}(M)$ has infinitely many path components detected by η, the same must be true for $\mathcal{R}_{scal}^{+}(M)/H$. But H has finite index in $\mathrm{Diff}(M)$ since there are only finitely many possible induced isomorphisms $\pi_1(M) \to \pi_1(M)$ by virtue of the fact that $\pi_1(M)$ is finite. It then follows that $\mathcal{R}_{scal}^{+}(M)/\mathrm{Diff}(M)$ has infinitely many path components.

The only remaining challenge is to find a class $[N,g',s',\sigma'] \in \Omega_{2k-1}^{Spin,+}(BG)$ for some finite group G and $\rho \in R_0(G)$, such that $\eta(N,g',s',\sigma')(\rho) \neq 0$. Botvinnik and Gilkey show that in any odd dimension at least five, and for any finite group G, such a class exists.

§8.2 *The work of Walsh*

Almost every investigation of the space of positive scalar curvature metrics has focused on connectedness issues, and the detection of non-triviality of homotopy groups. The aim in this section is to discuss a remarkable paper of Mark Walsh [Wa3], which looks at the space of positive scalar curvature metrics on S^n for $n \geq 3$. This paper studies topological issues for $\mathcal{R}^+_{scal}(S^n)$ which are more delicate than anything considered previously. These issues concern H-space and loop space structures, which we will now define.

A topological space X is an H-space if it is equipped with a continuous 'multiplication' map $\mu : X \times X \to X$, such that the following condition is fulfilled: there is an 'identity' element $e \in X$ such that the maps $x \mapsto \mu(x, e)$ and $x \mapsto \mu(e, x)$ are both homotopic to the identity map of X. Thus an H-space could be viewed as a homotopy notion analogous to a topological group. Note that there are other, stronger, notions of H-space, however for the spaces under consideration in the paper [Wa3], these definitions all coincide. There are concepts of homotopy commutativity and homotopy associativity for H-spaces. These are as expected: an H-space X is homotopy commutative if $\mu \simeq \mu \circ \omega$, where $\omega : X \times X \to X \times X$ is the 'flip' map $\omega(x, y) = (y, x)$. Similarly for homotopy associativity.

The loop space ΩX for a based space (X, x_0) the set of continuous loops in X based at x_0:

$$\Omega X := \{\gamma : [0,1] \to X \mid \gamma(0) = \gamma(1) = x_0\}.$$

This is then made into a topological space by giving it the compact-open topology. We can view ΩX as a based space by taking the constant loop at x_0 as base point. The loop space construction is easily iterated, each time taking the constant loop at the previous base point as the new base point. This results in so-called n-fold loop spaces $\Omega^n X = \Omega ... \Omega X$. In the sequel we will be interested in when a space Y is weakly homotopic to an n-fold loop spaces. This means that there is a continuous map $Y \to \Omega^n X$ for some space X, which induces isomorphisms on all homotopy groups.

It is not difficult to see that the concatenation of loops provides any loop space with an H-space structure. On the other hand, not every H-space is a loop space, nor is every H-space weakly equivalent to a loop space. Investigating when an H-space is (weakly equivalent to) a loop space is a delicate problem, and occupies a large proportion of the paper [Wa3] for the spaces under consideration there.

Let us now state the main results:

Theorem 8.2.1. *For $n \geq 3$, there is a subset $T \subset \mathcal{R}^+_{scal}(S^n)$ such that T admits an H-space structure and is homotopy equivalent to $\mathcal{R}^+_{scal}(S^n)$. Moreover, the H-space multiplication is both homotopy commutative and homotopy associative.*

Since $\mathcal{R}^+_{scal}(S^2)$ is known to be contractible ([RS]) and the fundamental groups of H-spaces are always abelian, we immediately have the following:

Corollary 8.2.2. *For any n, $\pi_1(\mathcal{R}^+_{scal}(S^n))$ is abelian.*

Turning our attention to loop spaces we have

Theorem 8.2.3. *For $n > 3$ the path component of $\mathcal{R}^+_{scal}(S^n)$ containing the round metric is weakly homotopy equivalent to an n-fold loop space.*

The above theorem raises the question of whether the entire space $\mathcal{R}^+_{scal}(S^n)$ is weakly homotopy equivalent to a loop space. In the case $n = 3$, it is known that $\mathcal{R}^+_{scal}(S^n)$ is path-connected (see [CM], or §8.4 in this chapter). Thus for the three-sphere, the space of positive scalar curvature metrics *is* weakly homotopy equivalent to a loop space. On the other hand, it is known that for $n \geq 5$, $\mathcal{R}^+_{scal}(S^n)$ has infinitely many path components (see §4.2 or §6.1). It turns out [Wa3; Theorem 9.3] that the loop space question can be answered in the affirmative provided a conjecture of Botvinnik [B1] (but see also [B2]) is true. This conjecture asserts that for $n \geq 5$, two positive scalar curvature metrics on the n-sphere are isotopic through positive scalar curvature metrics if and only if they are positive scalar curvature concordant.

Let us return to the first theorem above, which asserts that $\mathcal{R}^+_{scal}(S^n)$ has the homotopy type of a certain subset which admits an H-space structure. Although the proof contains many details which require careful handling, the basic idea is not difficult, and our next task is to describe this.

Fix a basepoint $p \in S^n$. The subset $T \subset \mathcal{R}^+_{scal}(S^n)$ is defined to be the set of positive scalar curvature metrics which take the form of a 'torpedo' in a neighbourhood of, and centred on the point p. A torpedo metric is a metric on a disc which is round near the centre of the disc, is globally rotationally symmetric about the centre, has non-negative sectional curvature, and is cylindrical at the boundary. A simple way to construct such a metric is to glue a hemisphere to a cylinder. This is not smooth, but it is easy to see that it can be smoothed so as to satisfy the requirements. In fact, for the purposes of the Walsh paper, it suffices to consider torpedoes which are only infinitesimally cylindrical at the boundary.

It can be shown ([Wa3; Lemma 4.7]) that $T \simeq \mathcal{R}^+_{scal}(S^n)$. Moreover, the same is true if we fix any finite number of basepoints and consider metrics which are torpedoes in a neighbourhood of all the basepoints.

There is an 'obvious' way in which to try and introduce a product operation into the space of positive scalar curvature metrics on the sphere: namely to perform a metric connected sum operation using the Gromov-Lawson procedure. However, in order to make this work, we need to select points in each sphere around which to perform the connected sum - hence the need to fix basepoints. We also encounter a potential problem due to the fact that the Gromov-Lawson process involves making choices. To rectify this, we can demand that in a neighbourhood of the basepoints, the metrics take a standard form, and in this way the Gromov-Lawson process can be made canonical by fixing choices once and for all. This explains the interest in torpedo metrics. Thus it might appear that we can introduce an H-space product operation into the subset T. However there is one further issue. We start with a pair of based spaces. After performing the connected sum, the resulting space is no longer based. This is a problem if we wish to use this object in further products.

To overcome the basepoint difficulty, Walsh introduces a 'connecting piece': a sphere with three basepoints p_0, p_1, p_2, equipped with a positive scalar curvature metric which is a torpedo in a neighbourhood of all the basepoints. Given an ordered pair (g_1, g_2) of elements of T, we form a Gromov-Lawson connected sum between g_i and the connecting piece in a neighbourhood of p_i, $i = 1, 2$. The resulting space still retains a basepoint, namely p_0. Using direct metric deformations, Walsh demonstrates that this product operation is both homotopy commutative and homotopy associative.

Let us now turn our attention to iterated loop spaces. It is convenient to introduce a new subset of $\mathcal{R}^+_{scal}(S^n)$: let $\mathcal{R}^+_{D_+(1)}(S^n)$ denote the set of positive scalar curvature metrics on the sphere for which a neighbourhood of the basepoint takes the form of a round unit radius hemisphere centered on the basepoint. Again we have that

$$\mathcal{R}^+_{scal}(S^n) \simeq \mathcal{R}^+_{D_+(1)}(S^n).$$

Establishing the loop space theorem cited at the top of this section hinges on showing that the connected component of the round metric in $\mathcal{R}^+_{D_+(1)}(S^n)$ admits a certain 'operad' action. Our next task is to outline this meaning of this term.

We will refrain from giving a definition of an operad: this is quite technical and not strictly necessary for our purposes. Instead, we will content ourselves by sketching the general idea. Roughly speaking, an operad is a countable collection of topological spaces $P_0, P_1, P_2, ...$ where each P_m admits an action of the symmetric group S_m, and where the initial space P_0 is a point. Together with this collection of spaces, there must also be continuous 'product' maps taking the form

$$P(k) \times P(j_1) \times ... \times P(j_k) \to P(j)$$

where $j = \sum_{i=1}^k j_i$, which satisfy a notion of associativity and a form of equivariance with respect to the symmetric group actions, and for which there is an element $1 \in P(1)$ which is a form of identity. Furthermore, there is a notion of 'action' for such an object on a topological space.

The most important example of an operad for present purposes is the so-called 'little discs operad'. The spaces which make up this object are defined as follows. Consider the standard unit radius round disc D^n. For fixed $m \in \mathbb{N}$, consider the set $P(m)$ of all possible configurations of i small disjoint open round discs embedded into D^n, where the radii of the 'little discs' are allowed to vary. This set of embeddings can be topologized as a subset of $(D^n \times [0,1])^m$, as each little disc is determined by a pair (p_s, ϵ_s) where $\epsilon_s < 1$ is the radius and p_s is the centre point, for $s = 1, ..., m$. Notice that each space $P(m)$ admits an action from the symmetric group S_m, which acts by permuting the order of the little discs (i.e. by permuting the subscripts). The product operation is similarly straightforward: given an element $\Delta \in P(k)$, i.e. a collection of k disjoint embedded little discs in D^n, after rescaling by suitable factors we can embed a collection of j_i little discs given by an element of $P(j_i)$ into the ith little disc of Δ, for each $i = 1, ..., k$. This results in a collection of $j := \sum_{i=1}^k j_i$ little discs embedded into the original D^n, which is an element of $P(j)$.

An operad action on a space X means a collection of actions $P(k) \times X^k \to X$ for each space $P(k)$ in the operad, subject to certain associativity and S_k-equivariance conditions.

The relevance of the little disc operad and operad actions for detecting iterated loop spaces is due to a result of Boardman, Vogt and May (see [BV], [Ma] and [V]). This states that if a path-connected space admits an action from the n-dimensional little discs operad, then X is weakly homotopy equivalent to an n-fold loop space. Thus in order to establish the second of the main theorems above, it suffices to show that the path component of $\mathcal{R}^+_{scal}(S^n)$ containing the round metric admits an action from the little discs operad.

It turns out to be sufficient to prove the existence of an action from a variant of the little discs operad, namely the operad which results from applying the so-called 'bar construction' to the little discs operad. The details of this will not concern us, however in the context of positive scalar curvature of metrics, establishing an action of this alternative operad is more natural. Indeed the definition of this action, which involves smoothly modifying the positive scalar curvature metric on S^n according to the particular operad element, is broadly similar in spirit to the metric modifications involved in the H-space multiplication operation. The details, however, both here and in many places in the paper [Wa3] are intricate, and require delicate and sometimes lengthy arguments.

We conclude this section by mentioning a result which appears in two different papers, one of which is authored by Walsh. The result in question is as follows:

Theorem 8.2.4. *([Wa2]) Suppose that M and N are closed simply connected n-manifolds, $n \geq 5$, which admit metrics of positive scalar curvature. If M and N are spin bordant, then $\mathcal{R}^+_{scal}(M) \simeq \mathcal{R}^+_{scal}(N)$. In particular, if M is a spin boundary, then $\mathcal{R}^+_{scal}(M) \simeq \mathcal{R}^+_{scal}(S^n)$.*

This theorem is an easy corollary of the following surgery result, using the fact that simply connected spin manifolds in dimensions greater than or equal to five are spin bordant if and only if they can be obtained from one another by surgeries of codimension at least three (see for example [St] for more details on this).

Theorem 8.2.5. *If a compact manifold Y^n, $n \geq 5$, is obtained from a manifold X by a surgery of codimension at least three, then $\mathcal{R}^+_{scal}(X) \simeq \mathcal{R}^+_{scal}(Y)$.*

This result originally appeared in the hitherto unpublished preprint of Chernysh [Ch]. It was published for the first time in Walsh's paper [Wa2] with a much shorter argument than that offered by Chernysh.

The approach taken by Walsh is as follows. As for the H-space and loop space considerations above, Walsh considers a special subset of $\mathcal{R}^+_{scal}(X)$ which consists of metrics which in some sense take a standard form near the sphere on which surgery is to be performed. Performing surgery according to the Gromov-Lawson technique then yields a positive scalar curvature metric on Y, which again is standard in the surgery neighbourhood, and allows for a complementary Gromov-Lawson surgery back to X. Thus we obtain maps between this locally standard set of positive scalar curvature metrics on X and that for Y. With suitable control over the various parameter choices needed for the construction, we can arrange for these maps to be homotopy equivalences. The challenge then is to show that the full space of positive scalar curvature metrics has the same homotopy type as the subset. The strategy for this is the same as that used in the analogous situations for the H-space and loop spaces structures. First we observe that by a result of Palais [Pa], all spaces of positive scalar curvature metrics under consideration are dominated by CW complexes. In other words, if S denotes one of the spaces of positive scalar curvature metrics, then there is a CW complex K and maps $S \to K \to S$ such that the composition is homotopic to the identity. By Whitehead's Theorem, it then suffices to show that the relative homotopy groups all vanish. This means that for any smooth family of positive scalar curvature metrics parametrized by a disc, with the metrics corresponding to the disc boundary belonging to the subset, we can simultaneously deform the metrics so that all the final metrics belong to the subset. That such a family deformation is possible was established in previous work of Walsh [Wal].

§8.3 *The work of Belegradek and Hu*

It is well-known that in dimension 2, up to constant factors the various measures of curvature - sectional, Ricci and scalar - all agree with the Gaussian curvature. Thus it suffices to speak of 'curvature' without any further qualification. In the compact case, Riemann's Uniformization Theorem provides the key to understanding spaces with curvature of a constant sign. This theorem states that for any given compact orientable surface M equipped with a metric g, there is a smooth function $u : M \to \mathbb{R}$ such that the conformally equivalent metric $e^{2u}g$ has constant curvature. It follows immediately from the Gauss-Bonnet Theorem that the sign of this constant curvature metric is an invariant of the surface. This allows us to give a complete description of the compact surfaces which admit curvatures of a given sign: the sphere and real projective space with positive curvature, the torus and Klein bottle with identically zero curvature, and all other surfaces with negative curvature.

If we use the fact that simply-connected Riemannian manifolds with constant (sectional) curvature are unique up to orientation preserving isometry (see [doC; page 163]), we can re-state the Uniformization Theorem in the following way: if M is a compact oriented surface then the group $C^\infty(M) \rtimes \mathrm{Diff}^+(M)$, where $\mathrm{Diff}^+(M)$ is the group of orientation preserving diffeomorphisms of M, acts transitively on the space of all Riemannian metrics on M via the action $(u, \phi) \cdot g = \phi^*(e^{2u}g)$.

Examining the curvature in more detail, suppose that g itself has curvature of constant *sign*, and that u is chosen such that e^{2u} has constant curvature. It then follows from the formula

$$scal(e^{2u}g) = e^{-2u}(scal(g) + 2\Delta u)$$

(see [Be; p. 59]) that the metrics $e^{2tu}g$ for $t \in [0, 1]$ all share the same curvature sign. Thus we see that each metric with everywhere positive (or negative) curvature belongs to the same path component of the space of metrics with positive (respectively negative) curvature as a constant curvature metric.

In some cases we can say more. It was shown in [RS] that the space of positively curved metrics on S^2 is actually contractible. In outline, the argument is as follows. From the second version of the Uniformization Theorem above, the space of all Riemannian metrics on S^2 can be identified with the quotient of $C^\infty(S^2) \rtimes \mathrm{Diff}^+(S^2)$ by the isotropy group of the round metric ds_2^2. Using classical results from complex analysis, this latter group can be identified with $PSL(2, \mathbb{C})$. By the above paragraph, the set $S \subset C^\infty(S^2)$ of scaling functions u for which $e^{2u}ds_2^2$ has positive curvature is a star-shaped set, and is therefore contractible. We deduce that the set of all positively curved metrics on S^2 can be identified with $S \rtimes \mathrm{Diff}^+(S^2)/PSL(2,\mathbb{C})$. As $C^\infty(S^2)$ and the space of all metrics $C^\infty(S^2) \rtimes \mathrm{Diff}^+(S^2)/PSL(2,\mathbb{C})$ are both contractible, we see that $\mathrm{Diff}^+(S^2)/PSL(2,\mathbb{C})$ is contractible. Using the contractibility of the set S, we conclude that the space of all positively curved metrics on S^2 is also contractible. Analogous arguments are presented in [RS] to show that the space of positively curved metrics on $\mathbb{R}P^2$ is similarly contractible. As a trivial consequence we deduce the path-connectedness of the corresponding moduli spaces.

The situation for non-compact manifolds, in contrast, is more complicated. The main aim of this section is to summarize the work of Belegradek and Hu in [BH]. This concerns metrics of non-negative curvature on $\mathbb{R}^2$. It turns out that the space of such objects has a remarkable structure.

The key to understanding such metrics goes back to the paper [BF] from 1942: a complete non-negatively curved metric g on $\mathbb{R}^2$ is isometric to $e^{-2u}g_0$, where g_0 is the standard Euclidean metric, and u is some smooth function. This means that g is equal (as opposed to merely isometric) to $\phi^* e^{-2u} g_0$ where ϕ is some self-diffeomorphism of $\mathbb{R}^2$. Moreover, it can be shown [BH; Lemma 3.1] that such an expression for g becomes unique if we demand that $\phi \in \mathrm{Diff}^+_{0,1}(\mathbb{R}^2)$, the group of self-diffeomorphisms of the plane which fix the complex numbers 0 and 1, and which are isotopic to the identity.

Let us consider the converse question of which metrics conformal to the Euclidean metric are complete with non-negative sectional curvature. The curvature condition has a straightforward characterization: the curvature of $e^{-2u}g_0$ is simply $e^{2u}\Delta u$, where Δ denotes the Laplacian for $(\mathbb{R}^2; g_0)$. Thus we obtain non-negative curvature for such a metric if and only if u is a subharmonic function. On the other hand, characterizing completeness is more difficult and is a central theme of the paper [BH]. It is possible to give a precise answer to this question, and in order to understand this we need the following quantity. For a function u, consider

$$\alpha(u) := \lim_{r\to\infty} \frac{M(r,u)}{\log r}$$

where

$$M(r,u) := \sup\{u(z) \mid |z| = r\}.$$

Now there is no guarantee that this limit makes sense, however it is well-known that if u is subharmonic then this limit exists in $[0,\infty]$. Careful consideration of subharmonic functions then leads to

Theorem 8.3.1. *([BH; Thm. 1.1]) The metric $e^{-2u}g_0$ is complete if and only if $\alpha(u) \leq 1$.*

If we define S_α to be the set of all smooth subharmonic functions on $\mathbb{R}^2$ with $\alpha(u) \leq \alpha$, we have that complete non-negatively curved metrics on $\mathbb{R}^2$ correspond to functions in S_1. Consequently we see there is a bijection between $S_1 \times \mathrm{Diff}^+_{0,1}(\mathbb{R}^2)$ and the set of all smooth, complete, non-negatively curved metrics on $\mathbb{R}^2$ given by mapping a pair (u,ϕ) to the metric $\phi^* e^{-2u} g_0$. Call this map Π.

We need to introduce a topology into the space of metrics. Belegradek and Hu use the symbol $\mathcal{R}^k_{\geq 0}(\mathbb{R}^2)$ to denote the space of smooth complete non-negatively curved metrics equipped with the C^k topology, and $\mathcal{R}^{k+s}_{\geq 0}(\mathbb{R}^2)$ to denote the same set with the C^{k+s} topology, $s \in (0, 1)$. Here we also allow $k = \infty$. Given such a topology, it is reasonable to ask whether the above bijection Π is actually a homeomorphism. The basic result is as follows:

Theorem 8.3.2. *([BH; Thm. 4.1]) If S_1 is equipped with the C^{k+s} topology and $\mathit{Diff}^+_{0,1}(\mathbb{R}^2)$ is given the C^{k+1+s} topology, then the map*

$$\Pi : S_1 \times \mathit{Diff}^+_{0,1}(\mathbb{R}^2) \to \mathcal{R}^k_{\geq 0}(\mathbb{R}^2)$$

is a homeomorphism. In particular this is the case when all sets are given the smooth topology.

Notice that the above result is no longer true if we equip the sets on the left-hand side with the smooth topology but use $\mathcal{R}^{k+s}_{\geq 0}(\mathbb{R}^2)$ on the right-hand side with k finite: it is easy to see that this map factors through the identity map $\mathcal{R}^{\infty}_{\geq 0}(\mathbb{R}^2) \to \mathcal{R}^{k+s}_{\geq 0}(\mathbb{R}^2)$, which is clearly a continuous bijection, but not a homeomorphism.

Central to the results of Belegradek and Hu are theorems from infinite dimensional topology. These results were proved elsewhere, and in many cases are classical. In order to understand the topology of $\mathcal{R}^{\infty}_{\geq 0}(\mathbb{R}^2)$, some of these results can be applied to the product $S_1 \times \mathrm{Diff}^+_{0,1}(\mathbb{R}^2)$ as follows. Any closed, convex, non-locally compact subset of a separable Fréchet space is homeomorphic to the sequence space l_2. In particular this applies to S_1 viewed as a subset of $C^{\infty}(\mathbb{R}^2)$. The space $\mathrm{Diff}^+_{0,1}(\mathbb{R}^2)$ is also known to be homeomorphic to l_2, and furthermore $l_2 \times l_2$ is again homeomorphic to l_2. This yields:

Theorem 8.3.3. *([BH; Thm. 1.2]) $\mathcal{R}^{\infty}_{\geq 0}(\mathbb{R}^2)$ is homeomorphic to l_2.*

Having identified the topological type of $\mathcal{R}^{\infty}_{\geq 0}(\mathbb{R}^2)$, it still remains to say something about the topology of $\mathcal{R}^k_{\geq 0}(\mathbb{R}^2)$ for finite k. The identity map

$$\mathcal{R}^{\infty}_{\geq 0}(\mathbb{R}^2) \to \mathcal{R}^k_{\geq 0}(\mathbb{R}^2)$$

is a continuous bijection, and as an injective map to a Hausdorff space, it restricts to a homeomorphism on every compact subset. An example of such a compact subset is given by the Hilbert cube: this is the infinite product $\Pi^{\infty}_{n=1}[0, 1/n]$ equipped with the product topology. As a product of compact intervals, the Hilbert cube is clearly compact. Moreover, since l_2 is homeomorphic to the countably infinite product $(-1, 1)^{\mathbb{N}}$, we see that the Hilbert cube is embeddable in l_2. Denoting the Hilbert cube by H, we therefore have a composition of embeddings

$$H \to l_2 \cong \mathcal{R}^{\infty}_{\geq 0}(\mathbb{R}^2) \to \mathcal{R}^k_{\geq 0}(\mathbb{R}^2).$$

Using the classical facts that every separable metrizable space embeds into the Hilbert cube, and the complement in l_2 of a countable union of compact subsets is again homeomorphic to l_2, the following intriguing connectedness result is easily deduced:

Theorem 8.3.4. *([BH; Thm. 1.4]) If K is a countable subset of $\mathcal{R}^k_{\geq 0}(\mathbb{R}^2)$ and X is any separable metrizable space, then for any distinct points $x_1, x_2 \in X$ and any distinct metrics $g_1, g_2 \in \mathcal{R}^k_{\geq 0}(\mathbb{R}^2) \setminus K$, there is an embedding of X into $\mathcal{R}^k_{\geq 0}(\mathbb{R}^2) \setminus K$ which maps x_i to g_i, $i = 1, 2$.*

Among other results in their paper, Belegradek and Hu also comment on the topology of the moduli space of non-negatively curved metrics $\mathcal{M}^k_{\geq 0}(\mathbb{R}^2)$:

Theorem 8.3.5. *([BH; Thm. 1.6]) The complement of a subset S of $\mathcal{M}^k_{\geq 0}(\mathbb{R}^2)$ is path connected if S is countable, or if S is closed and finite dimensional.*

It should be noted that $\mathcal{M}^k_{\geq 0}(\mathbb{R}^2)$ is not Hausdorff. Indeed a non-flat metric g can be found in $\mathcal{R}^k_{\geq 0}(\mathbb{R}^2)$ ([BH; Proposition 3.4]) for which every neighbourhood of every metric isometric to the Euclidean metric g_0 contains a metric isometric to g.

§8.4 *The work of Codá Marques*

In §4.2 we saw that Carr proved that the space of positive scalar curvature metrics on spheres S^{4n-1} has infinitely many path components when $n \geq 2$. That the same conclusion also holds for the moduli space follows from the work of Kreck and Stolz discussed in §6.1. This raises the question of what happens in the case $n = 1$, i.e. for S^3, or more generally for any compact orientable three-manifold. As discussed in §8.3, the space of positive scalar curvature metrics on S^2 is contractible ([RS]).

In [CM], Codá Marques proves the following results:

Theorem 8.4.1. *If M is a compact orientable three-manifold which admits a positive scalar curvature metric, then the moduli space of positive scalar curvature metrics on M is path-connected.*

In [Ce1] it was shown that the space of oriented diffeomorphisms of S^3 is also path-connected. Combining this with the above theorem immediately gives:

Corollary 8.4.2. *The space of positive scalar curvature metrics on S^3 is path-connected.*

We will outline the main features of the proof of the above theorem, however the paper [CM] contains a more detailed overview.

The proof depends on Ricci flow techniques, and in particular on Perelman's work. Recall that given an initial Riemannian metric g_0 on a manifold M, the Ricci flow is the solution to the initial value problem

$$\frac{\partial g}{\partial t} = -2Ric(g),$$

where $g = g(x,t)$ is a one-parameter family of Riemannian metrics on M subject to the initial condition $g(x,0) = g_0(x)$ for all $x \in M$. It is not difficult to show that if g_0 has positive scalar curvature then the flow must terminate in finite time. Indeed we can write down a positive lower bound for the scalar curvature which tends to infinity at some finite point in time, which reflects the formation of singularities in some (typically proper) subset of the manifold. There is a notion of surgery for the Ricci flow which removes singular neighbourhoods and replaces them with some geometrically standard piece, which then allows the flow to be re-started.

It follows from Perelman's work on Ricci flow with surgeries that a compact orientable three-manifold which admits a positive scalar curvature metric must be diffeomorphic to a connected sum of finitely many spherical space forms (i.e. quotients S^3/Γ where $\Gamma \subset \mathrm{SO}(4)$ is a finite group acting freely on S^3) and copies of $S^2 \times S^1$. It therefore suffices to focus on manifolds of this type.

Codá Marques introduces the concept of a 'canonical metric' for a manifold

$$S^3 \sharp (S^3/\Gamma_1) \sharp \cdots \sharp (S^3/\Gamma_k) \sharp (S^2 \times S^1) \sharp \cdots \sharp (S^2 \times S^1).$$

This is constructed as follows. Take a round metric of radius one on S^3, and metrics of constant curvature one on each of the spherical space forms. To extend these metrics across the connected sums, use the Gromov-Lawson technique. Moreover, a positive scalar curvature metric on each connected sum with $S^2 \times S^1$ can be achieved by performing a Gromov Lawson connected sum of S^3 to itself. Clearly there are many choices involved in the construction of a canonical metric, however we can say the following:

Lemma 8.4.3. *All canonical metrics on M lie in the same path component of the moduli space $\mathcal{R}^+_{scal}(M)/Diff(M)$.*

The proof of this lemma relies in part on work of de Rham [deR] and Milnor [M1] which shows that given two canonical metrics for M, in the corresponding connected sum decompositions the spherical space forms correspond pairwise, and moreover there is an orientation preserving isometry between each pair. For this reason the conclusion of the lemma (and in turn the main theorem) involves moduli spaces instead of spaces of positive scalar curvature metrics. We also need to consider the various choices required to construct the connected sums. The first choice involves the locations at which the connected sum 'tube' is to be connected to the manifold(s). These locations can be determined by a choice of two points, with the ends of the tube meeting the boundaries created when small discs centered on the two points are removed. The second choice involves selecting an orthonormal tangent frame at these points (compatible with the orientation), as this is the first step in Gromov and Lawson's argument. It is not difficult to see that two different choices of points can be linked along smooth paths, and a smoothly varying choice of orthonormal bases can be chosen along the path linking any two given bases at the endpoints. Thus the Gromov-Lawson metrics which result from different choices of points/frames can be connected along a one parameter family of Gromov-Lawson metrics. Hence given two canonical metrics on M, after pulling one of them back by a suitable self-diffeomorphism of M the resulting metrics must be isotopic, which establishes the lemma.

In the light of the above lemma, to prove the main theorem it suffices to show that any positive scalar curvature metric on a compact orientable three-manifold is isotopic to a canonical metric.

The idea is to subject the given metric to the Ricci flow with surgeries. Given that the initial metric has positive scalar curvature, the Ricci flow with surgeries becomes extinct in finite time. In other words, at some point the Riemannian manifold given by the flow with surgeries becomes the empty set. Just before the extinction time, each connected component of the manifold (and there might be several resulting from previous singularity formation/surgeries) has extremely large (positive) scalar curvature. One consequence of the appearance of high scalar curvature regions in the flow is that all points in such regions can be shown to belong to one of four possible standard neighbourhoods. A careful analysis of such neighbourhoods shows that if every point in a compact orientable Riemannian three-manifold belongs to one of these, then the metric on the manifold must be isotopic to a canonical metric. Thus just before the extinction time for the flow with surgeries, each component of our manifold has a canonical metric up to isotopy.

To complete the proof of the main theorem requires a backward induction argument. This argument is roughly as follows. As the flow preserves the isotopy type of the metric, we only need focus on the times just before and just after a surgery. By Perelman's work, each component of the manifold just before a surgery is a connected sum of the components after the surgery, together with connected sums with certain other manifolds which have uniformly high scalar curvature. These latter manifolds, which by the above paragraph have metrics isotopic to canonical metrics, are removed by the surgery process. Thus if each component after the surgery has a canonical metric up to isotopy, the same must be true for those before the surgery, subject to showing that a Gromov-Lawson connected sum of canonical metrics is isotopic to a canonical metric. The theorem then follows by backward induction over the singular times at which surgery is performed.

§8.5 *The work of Lohkamp*

Although this book is largely concerned with (moduli) spaces of various kinds of positively curved metrics, it is worth considering the case of negative scalar and negative Ricci curvature metrics. These were investigated by Lohkamp. (Moduli spaces of negative sectional curvature metrics will be treated in detail in the next chapter.) Lohkamp's results provide some context for their positively curved analogues, and also act as a counterpoint to them, as they demonstrate the asymmetry between the positive and the negative in spectacular fashion. Indeed Lohkamp's results about spaces of negative Ricci or scalar curvature metrics are strikingly simple (though their proofs are not). In particular, complete answers can be given to the fundamental questions.

At the heart of the difference between positive and negative Ricci or scalar curvature is the fact that the negative curvatures display a certain local flexibility not shared by the corresponding positive curvatures. As an example of this, consider the following

Theorem 8.5.1. *([L3; Theorem C]) Let M be a manifold of dimension at least 3, (either compact or non-compact), and let g_i, $i = 0, 1$, be metrics with Ricci curvature $Ric(g_i) < \alpha$ for some $\alpha \in \mathbb{R}$. Suppose that we have a closed set S and an open subset U with $S \subset U \subset M$, such that g_0 and g_1 agree on $M \setminus S$. Then there is a continuous family of metrics g_t, $t \in [0, 1]$, such that $Ric(g_t) < \alpha$ for all t and $g_t = g_0 = g_1$ on $M \setminus U$. The analogous result holds for scalar curvature.*

It is important to note that the spaces of negative Ricci and scalar curvature metrics are *not* convex. The metric deformations required to prove the above result are much more subtle.

At many points in this book we have seen examples of manifolds for which the space of positive scalar curvature metrics is not path connected. It is clear that the positive scalar curvature analogue of the above theorem cannot hold if we apply it to metrics in different path components.

Perhaps the most fundamental result about negative Ricci or scalar curvature metrics is the following:

Theorem 8.5.2. *([L1], [L2]) Let M^n be any smooth manifold of dimension $n \geq 3$, either compact or non-compact. Then M admits a complete metric of negative Ricci curvature. Moreover (i) there are constants $a(n) > b(n) > 0$ such that there is a metric g on M with $-a(n) < Ric(g) < -b(n)$; (ii) if M is non-compact ($n \geq 2$) then for any metric g_0 on M, there is a $Ric < 0$ metric in the same conformal class (i.e. $e^{2f} g_0$ for some $f : M \to \mathbb{R}$).*

It is a trivial observation that the analogous result must hold for negative scalar curvature.

Corollary 8.5.3. *There are no topological obstructions to negative Ricci or scalar curvature in dimensions at least 3.*

The idea behind establishing the existence of negative Ricci curvature metrics on an arbitrary manifold is very roughly as follows. On $\mathbb{R}^n$ it can be shown that there is a metric g such that $Ric(g) < 0$ on the open unit ball $B_1(0)$ about the origin, and $g = g_{Eucl}$ on $\mathbb{R}^n \setminus B_1(0)$. Now on a small enough scale, any Riemannian manifold 'looks' flat. (Given any geodesic ball $B_\epsilon(x)$, identifying T_xM isometrically with Euclidean space and using the exponential map composed with the obvious scaling map, pull back the metric on $B_\epsilon(x)$ to $B_1(0) \subset \mathbb{R}^n$. The C^k-difference between the pull-back and standard metric on $B_1(0)$ goes to 0 as $\epsilon \to 0$.) Combining the above observations, we can locally make the Ricci curvature more negative. Working with respect to a carefully chosen covering and repeating in all neighbourhoods eventually gives a global negative Ricci curvature metric.

If we turn our attention to spaces of metrics, we have the following:

Theorem 8.5.4. *([L3]) The set of metrics for which the Ricci curvature is uniformly bounded above by some constant $\alpha \in \mathbb{R}$ is contractible in the smooth topology. The same is true for the scalar curvature.*

Corollary 8.5.5. *The moduli space of negative Ricci or scalar curvature metrics is connected.*

The idea behind this theorem is as follows. For simplicity let us focus on the case $\alpha = 0$, i.e. on the set of negative Ricci or negative scalar curvature metrics. Start with a continuous map from the sphere S^i into the subset of negatively curved metrics, and then extend to a continuous map from D^{i+1} into the space of all metrics. Next, perform local deformations of the metrics in the image so as to coherently move the image, fixing the boundary, into the space of negative Ricci or negative scalar curvature metrics. This establishes the vanishing of all homotopy groups for the space of metrics. A result of Palais asserts that our space of metrics is dominated by a CW complex, and when combined with Whitehead's Theorem this allows us to deduce contractibility. (Compare §8.2.)

§9. Moduli spaces of Riemannian metrics with negative sectional curvature

In this chapter we shall discuss moduli spaces of negatively curved metrics. Here, in fact, only the case of *sectional* curvature is of any further interest as a consequence of the work of Lohkamp discussed in §8.5.

When studying moduli spaces of metrics with negative sectional curvature we will adopt a slightly different viewpoint. This is forced upon us because of the results of Teichmüller theory and Mostow's rigidity theorem, which we will now explain.

Let $S := S_g$ be a closed and, for simplicity, orientable topological surface of genus g. Recall that a complex atlas on S is a topological atlas for which all coordinate changes are biholomorphic maps. Two complex atlases on S are called equivalent if their union is also a complex atlas, and a maximal atlas inside such an equivalence class is called a *complex structure*. Equipping S with a complex structure makes S into a *Riemann surface*.

If we let $\mathcal{C} := \mathcal{C}_S := \{C \,|\, C \text{ is a complex structure on } S\}$ denote the set of all complex structures on the surface S, then $\mathrm{Homeo}(S)$, the group of self-homeomorphisms of S, acts on $\mathcal{C}$ in a natural way as follows. If $C \in \mathcal{C}$ and $\varphi : U \to \mathbb{C}$ is a chart of C, then for $f \in \mathrm{Homeo}(S)$, $\varphi \mapsto \varphi \circ f$ defines an action $\mathrm{Homeo}(S) \times \mathcal{C}_S \to \mathcal{C}_S$, whose orbit space is called the *Riemann moduli space* $\mathcal{M}(S)$ of S. Moreover, the normal subgroup $\mathrm{Homeo}_0(S) \triangleleft \mathrm{Homeo}(S)$ of self-homeomorphisms of S which are isotopic to the identity mapping also acts on $\mathcal{C}$, and the corresponding orbit space $\mathcal{C}_S/\mathrm{Homeo}_0(S)$ is called the *Teichmüller space* $\mathcal{T}(S)$. In particular, the so-called *mapping class group* $\mathrm{Homeo}(S)/\mathrm{Homeo}_0(S)$ then acts on $\mathcal{T}(S)$ with quotient $\mathcal{M}(S)$.

It turns out that the Teichmüller space $\mathcal{T}(S)$ is itself always a complex manifold. In particular, $\mathcal{T}(S^2)$ is a point, $\mathcal{T}(T^2)$ is the upper complex halfplane, and $\mathcal{T}(S_g) = \mathbb{C}^{3g-3}$ for $g > 1$, so $\mathcal{T}(S)$ is therefore always contractible.

Complex structures and Teichmüller space can also be understood from a more differential geometric perspective. Recall that two Riemannian metrics on a smooth manifold M are called *conformally equivalent* if one can be obtained from the other by multiplying by a smooth function $M \to \mathbb{R}^+$. Such metrics are said to belong to the same conformal class.

For a real two-dimensional orientable smooth surface S, it is well known that the conformal equivalence classes of Riemannian metrics on S are in one to one correspondence with the complex structures on S (compare, e.g., [BoKl], p.10). Moreover, by the Uniformization Theorem, for genus $g > 1$ any conformal class of metrics on S_g contains a uniquely determined hyperbolic structure, that is, a metric with constant sectional curvature -1. Thus, in this case, $\mathcal{T}(S)$ is simply the space of distinct hyberbolic structures on S, i.e.

$$\mathcal{T}(S) = \{(X, f)\}/\sim$$

where X is S equipped with a hyperbolic metric, $f : S \to X$ is a homeomorphism, and

$$(X, f) \sim (Y, g) \iff \exists \text{ isometry } i : X \to Y \text{ such that } i \circ f \text{ is homotopic to } g.$$

In higher dimensions, however, the analogous concept is not interesting, and this is due to the following result:

W. Tuschmann, D.J. Wraith, *Moduli Spaces of Riemannian Metrics*,
Oberwolfach Seminars 46, DOI 10.1007/978-3-0348-0948-1_9

Theorem 9.1. *(Mostow's Rigidity Theorem [Mos].) Let M^n and N^n be complete finite volume hyperbolic n-manifolds, where $n \geq 3$. Then any homotopy equivalence between M and N is homotopic to an isometry.*

Consequently, in dimensions at least three the standard Teichmüller space is just a point. In order to work with an object which is interesting in higher dimensions but still similar in spirit to Teichmüller space, we proceed as follows.

Consider again an oriented surface $S = S_g$ with genus $g > 1$. Using the Ricci flow one can show that any metric on S can be canonically deformed into a hyperbolic one. Moreover, any negatively curved metric on S will be deformed *through* negatively curved metrics into a hyperbolic one. Thus, the space of hyperbolic metrics on S is in a canonical way a deformation retract of the space of all negatively curved metrics on S.

Let $\mathrm{Diff}(S)$ denote the group of self-diffeomorphisms of S, and let $\mathrm{Diff}_0(S) \triangleleft \mathrm{Diff}(S)$ be the subgroup consisting of all self-diffeomorphisms of S that are homotopic to the identity. The Ricci flow commutes with the action of the group $\mathrm{Diff}(S)$, and thus the Teichmüller space $\mathcal{T}(S)$ is canonically a deformation retract of the quotient of the space of all negatively curved metrics on S by the action of $\mathrm{Diff}_0(S)$. Notice also that instead of considering the space of all negatively curved metrics on S, here we could also consider the space of all negatively pinched metrics on S, or even just the space of all Riemannian metrics on S.

(For foundational material and basic results on the Ricci flow we recommend the book [CLN].)

We are thus led to the following new setting. For a closed smooth manifold M which admits a negative sectional curvature metric and $\mathrm{Diff}_0(M) \triangleleft \mathrm{Diff}(M)$ as above, let

$$\mathcal{D}(M) := \mathbb{R}^+ \times \mathrm{Diff}(M).$$

The group $\mathcal{D}(M)$ then acts by scaling and pushing forward on the space $\mathcal{R}(M)$ of all Riemannian metrics on M via

$$(\lambda, \phi) \cdot g = \lambda \cdot (\phi^{-1})^*(g) = \lambda \cdot \phi_*(g),$$

and we define

$$\mathcal{M}(M) := \mathcal{R}(M)/\mathcal{D}(M)$$

to be *the moduli space of Riemannian metrics on M*. Moreover, $\mathcal{D}_0(M) := \mathbb{R}^+ \times \mathrm{Diff}_0(M)$ then also acts on $\mathcal{R}(M)$, and we denote by

$$\mathcal{T}(M) := \mathcal{R}(M)/\mathcal{D}_0(M)$$

the *Teichmüller space of Riemannian metrics on M*.

Now for $0 \leq \varepsilon \leq \infty$, let $\mathcal{R}^\varepsilon(M)$ be the space of all ε-pinched negatively curved metrics on M, i.e., the space of all Riemannian metrics g on M which satisfy the inequality

$$\frac{\sup \sec_g}{\inf \sec_g} \leq 1 + \varepsilon.$$

Thus, $g \in \mathcal{R}^\varepsilon \iff \exists \lambda \in \mathbb{R}^+ : -(1+\varepsilon) \leq \sec_{\lambda g} \leq -1$, so in particular $\mathcal{R}^0(M)$ is the space of all metrics on M with constant negative sectional curvature, and $\mathcal{R}^\infty(M)$ denotes the space of all negatively curved metrics on M. By analogy with the above, we also have a moduli space

$$\mathcal{M}^\varepsilon(M) := \mathcal{R}^\varepsilon(M)/\mathcal{D}(M),$$

as well as a Teichmüller space

$$\mathcal{T}^\varepsilon(M) := \mathcal{R}^\varepsilon(M)/\mathcal{D}_0(M)$$

of ε-pinched negatively curved Riemannian metrics on M.

Remarks 9.2.

1.) For a closed surface S_g as above with genus $g > 1$, the original Teichmüller space $\mathcal{T}(S_g)$ is now just $\mathcal{T}^0(S_g)$, and $\mathcal{T}^0$, $\mathcal{T}^\varepsilon$, $\mathcal{T}^\infty$ are all contractible.

2.) If M is closed hyperbolic and $\dim M \geq 3$, then by Mostow Rigidity, $\mathcal{T}^0(M)$ is a point, and so $\mathcal{R}^0(M) \cong \mathcal{D}_0(M)$.

3.) If M is closed hyperbolic, then since M is aspherical and $\pi_1(M)$ is centerless, $\mathcal{D}_0(M)$ acts freely on $\mathcal{R}(M)$. Moreover, $\mathcal{R}(M)$ is contractible, and

$$\mathcal{D}_0(M) \to \mathcal{R}(M) \to \mathcal{T}(M)$$

is a principal bundle, so $\mathcal{T}(M)$ is a classifying space $B\mathcal{D}_0(M)$ of the group $\mathcal{D}_0(M)$.

All things considered, for a hyperbolic closed manifold M we arrive at the following commutative diagram:

$$\begin{array}{ccccccc}
\mathcal{D}_0(M) \simeq \mathcal{R}^0(M) & \hookrightarrow & \mathcal{R}^\epsilon(M) & \hookrightarrow & \mathcal{R}^\infty(M) & \hookrightarrow & \mathcal{R}(M) \simeq * \\
\downarrow & & \downarrow & & \downarrow & & \downarrow \\
* \simeq \mathcal{T}^0(M) & \hookrightarrow & \mathcal{T}^\epsilon(M) & \hookrightarrow & \mathcal{T}^\infty(M) & \hookrightarrow & \mathcal{T}(M) = B\mathcal{D}_0(M) \\
\downarrow & & \downarrow & & \downarrow & & \downarrow \\
\mathcal{M}^0(M) & \hookrightarrow & \mathcal{M}^\epsilon(M) & \hookrightarrow & \mathcal{M}^\infty(M) & \hookrightarrow & \mathcal{M}(M).
\end{array}$$

Farrell and Ontaneda have thoroughly investigated spaces of negatively curved metrics. We will now state several of their results. Firstly, stated informally, we have the following:

Theorem 9.3. *([FO1]) The maps $\mathcal{T}^\varepsilon(M) \hookrightarrow \mathcal{T}^\infty(M)$ and $\mathcal{T}^\infty(M) \to \mathcal{T}(M)$ are in general not homotopic to constant maps, so that for $0 \leq \varepsilon \leq \infty$, the spaces $\mathcal{T}^\varepsilon(M)$ are in general not contractible.*

More precisely, one has the following statement:

Theorem 9.4. *([FO1]) For all $k_0 \in \mathbb{N}$, $n \geq 1$, there exists $n_0 = n_0(k) \in \mathbb{N}$ with the following properties: given $\varepsilon > 0$ and a closed real hyperbolic n-manifold M, where $n \geq n_0$, there exists a finite cover N of M such that for all $1 \leq k \leq k_0$ with $n + k \equiv 3 \bmod 4$, the map $\pi_k(\mathcal{T}^\varepsilon(N)) \to \pi_k(\mathcal{T}(N))$ induced by inclusion is nonzero. Hence, $\pi_k(\mathcal{T}^\varepsilon(N)) \neq 0$, and in particular $\mathcal{T}^\delta(N)$ is not contractible for every $\varepsilon \leq \delta \leq \infty$ (provided $k_0 \geq 4$).*

In addition, one can choose $N = M$ if M is a π-manifold (that is, if M^n embeds in $\mathbb{R}^{2n+2}$ with trivial normal bundle) and if the injectivity radius at some point of M is sufficiently large (depending only on M).

Moreover, Farrell and Ontaneda also proved:

Theorem 9.5. *([FO2]) If M is a closed hyperbolic n-manifold with a "good enough" closed geodesic, then $\pi_k(\mathcal{R}^-_{sec}(M))$ and $\pi_k(\mathcal{R}^-_{sec}(M)/(\mathbb{R}^+ \times \mathrm{Diff}(M)))$ are nontrivial, provided that either*

(1) $k = 0$, $n \geq 10$;
(2) $k = 1$, $n \geq 12$;
(3) $k = 2p - 4$, where $p > 2$ is a prime and $n \geq 3k + 8$.

Note that the property of being "good enough" depends on n and k, and can be achieved by taking finite covers N of nonarithmetic hyperbolic manifolds M.

Finally, we would also like to mention the following result:

Theorem 9.6. *([FO3]) Let M be a closed negatively curved n-manifold, where $n \geq 10$. Then*

(1) $\mathcal{R}^-_{sec}(M)$ has infinitely many path components;
(2) for all primes $2 < p < \frac{n+5}{6}$ and all real $\varepsilon > 0$, the homotopy groups $\pi_{2p-4}(\mathcal{R}_{sec<\varepsilon}(M))$ are nontrivial.

§10 Non-negative sectional curvature moduli spaces on open manifolds

In this chapter we will describe results about spaces and moduli spaces of complete Riemannian metrics with non-negative sectional curvature on open manifolds. A new and important tool for understanding these spaces involves employing properties of the so-called 'souls' of the metrics, and we start with a discussion of these.

§10.1 *Open non-negatively curved manifolds and their souls*

One of the fundamental structure results for open manifolds with non-negative sectional curvature is the following result of Cheeger and Gromoll:

Theorem 10.1.1. *(The Soul Theorem [CG1]) Let $M = (M, g)$ be an open (i.e., connected, non-compact, without boundary) complete Riemannian manifold with non-negative sectional curvature. Then there exists a closed totally geodesic and totally convex submanifold S, called the soul of (M, g), such that M is diffeomorphic to the normal bundle of S.*

Recall that $S \subset (M, g)$ is called *totally convex* if, for any two points, any geodesic connecting these points is completely contained in S, and called *totally geodesic* if any geodesic of S is also a geodesic of M.

Perelman later proved the following long-standing conjecture.

Theorem 10.1.2. *(Cheeger-Gromoll's Soul Conjecture [Per]) Let $M = M^n$ be as in the Soul Theorem, and suppose in addition that there exists at least one point in M where all sectional curvatures are strictly positive. Then the soul of M is a point, and hence M^n is diffeomorphic to $\mathbb{R}^n$.*

To a large extent the Soul Theorem reduces the study of complete open manifolds of non-negative sectional curvature to that of the closed case, since many of the topological properties of such manifolds are reflected in the soul.

Let us look at some examples.

Examples 10.1.3.

- The paraboloid $P = \{z = x^2 + y^2 \mid (x, y, z) \in \mathbb{R}^3\}$ is, when viewed as a subspace of Euclidean space, positively curved and has a unique soul, namely the point at its tip.

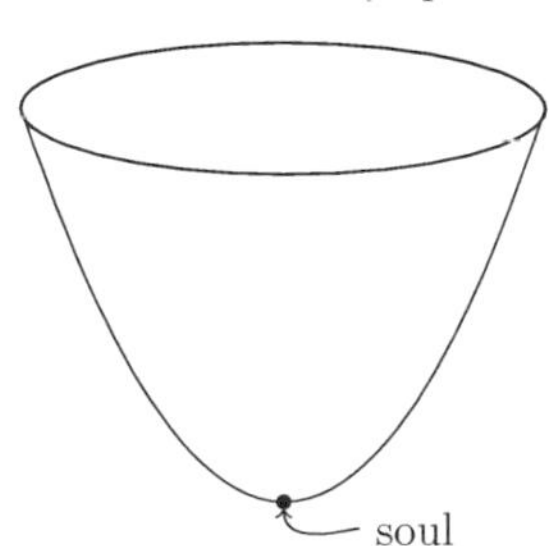

W. Tuschmann, D.J. Wraith, *Moduli Spaces of Riemannian Metrics*, Oberwolfach Seminars 46, DOI 10.1007/978-3-0348-0948-1_10

- For the flat cylinder $S^1 \times \mathbb{R} \subset \mathbb{R}^3$, any 'vertical' circle is a soul.

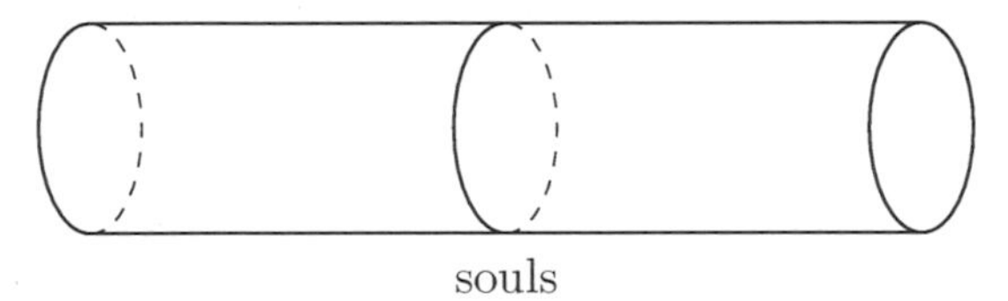

- If M is flat Euclidean n-space, then any point of M is a soul.

In particular, souls are not necessarily unique. However there is also some rigidity as shown by the following results of Sharafutdinov.

Theorem 10.1.4. *([Sha1], [Sha2]) For any open non-negatively curved manifold M with a soul S,*

1) *there exists a distance-nonincreasing retraction $\pi : M \to S$;*
2) *any two souls of M are isometric and can be mapped to each other by a diffeomorphism of the ambient space M.*

The regularity properties of the above Sharafutdinov retraction map π were subsequently improved as follows. In 1994, Perelman proved that π is a C^1 Riemannian submersion, Berestovski and Guijarro then improved this to $C^{1,1}$, and Guijarro further improved this to C^2 in 2000. Finally, in 2007 Wilking showed that π is in fact C^∞. (See [Per], [BeGu], [Gui], [Wil].)

Let us now begin our investigation of moduli spaces of non-negatively curved metrics on open manifolds. Suppose that we are given two different non-negatively curved metrics on the same open manifold M. What can be said about their respective souls? Should we expect these to be different in general?

A crucial topological tool for further constructions is given by Theorem 10.1.6 below, which follows from a combination of results of Haefliger and Siebenmann. In order to state this we first need a definition.

Definition 10.1.5. *A homotopy equivalence of smooth manifolds $f : X^n \to Y^n$ is a tangential homotopy equivalence if the vector bundles f^*TY and TX are stably isomorphic.*

This condition is *much* stronger than merely being homotopy equivalent! For example in a fixed homotopy type one can have an infinite number of tangential homotopy types. Such examples can, for instance, be found in the class of hyperbolic manifolds.

Theorem 10.1.6. *('Work Horse' Theorem [Hae], [Sie]) Let $\mathbb{R}^l \to E_i \to N_i^k$, $i = 1, 2$, be two vector bundles over smooth closed manifolds N_i^k, and suppose that $f : E_1 \to E_2$ is a tangential homotopy equivalence, where $l \geq 3$ and $l > k$. Then f is homotopic to a diffeomorphism.*

Let us now look at two examples. We begin with lens spaces.

Let p, q be coprime integers and let $S^3 \subset \mathbb{C}^2$ be equipped with the free $\mathbb{Z}_p$ action generated by

$$(z_1, z_2) \mapsto (e^{\frac{2\pi i}{p}} z_1, e^{\frac{2\pi i q}{p}} z_2).$$

Then the corresponding quotient manifold $S^3/\mathbb{Z}_p$ is called a lens space $L(p;q)$.

It is a classic fact that $L(p;q_1)$ and $L(p;q_2)$ are

- homotopy equivalent if and only if $q_1q_2 \equiv \pm n^2 \bmod p$ for some $n \in \mathbb{N}$;
- homeomorphic if and only if either $q_1 \equiv \pm q_2 \bmod p$ or $q_1q_2 \equiv \pm 1 \bmod p$.

Consider for example the lens spaces $L(7;1)$ and $L(7;2)$. These spaces are homotopy equivalent but not homeomorphic. However, since orientable three-manifolds are parallelizable, any homotopy equivalence between them is tangential. Define $E_1 := L(7;1) \times \mathbb{R}^4$ and $E_2 := L(7;2) \times \mathbb{R}^4$, and equip both manifolds with the obvious non-negatively curved product metrics. Applying the Work Horse Theorem then yields an open seven-manifold with two different non-negatively curved metrics whose souls are non-homeomorphic.

Our second example involves exotic spheres.

All exotic 7-spheres are stably parallelizable, and therefore by the Work Horse Theorem they all become diffeomorphic after taking products with $\mathbb{R}^8$ (as a matter of fact, in this case $\mathbb{R}^5$ already suffices).

Grove and Ziller [GZ] showed that any Milnor sphere (that is, a smooth homotopy 7-sphere that is diffeomorphic to an S^3-bundle over S^4) admits a Riemannian metric with non-negative sectional curvature. Thus, for any of the ten Milnor 7-spheres Σ^7, the manifold $S^7 \times \mathbb{R}^8$ admits a complete non-negatively curved Riemannian metric with soul diffeomorphic to Σ.

This example also indicates that the classification of non-negatively curved metrics on $S^k \times \mathbb{R}^l$ will in general be a hard problem.

The lens space and exotic sphere examples discussed above display only a finite number of non-negatively curved metrics with non-diffeomorphic souls on a fixed manifold. However, as Belegradek [Bel] showed in 2003, one can also find examples which exhibit an infinite number:

Theorem 10.1.7. *([Bel]) On the manifold $S^3 \times S^4 \times \mathbb{R}^5$ there exist infinitely many complete non-negatively curved Riemannian metrics with pairwise non-homeomorphic souls.*

The idea of the proof is as follows. Belegradek considers a certain collection of S^3-bundles over S^4 which are pairwise non-homeomorphic. These then are used as the base spaces for a corresponding collection of $\mathbb{R}^5$-bundles, which can also be viewed as $S^3 \times \mathbb{R}^5$-bundles associated to $\mathrm{SO}(3) \times \mathrm{SO}(3)$-principal bundles over S^4. By [GZ; Theorem F] such principal bundles admit invariant metrics of non-negative curvature, and hence by the O'Neill formulas for Riemannian submersions, so do the associated $S^3 \times \mathbb{R}^5$-bundles. Moreover the souls of these bundles are precisely the original S^3-bundles over S^4. Using some algebraic topology, Belegradek shows that for his choice of bundles, each total space is tangentially homotopy equivalent to $S^3 \times S^4 \times \mathbb{R}^5$. The result then follows from the Work Horse Theorem.

After Belegradek's result, one may also ask whether it is possible to obtain examples with better geometric properties and control of the souls, for example, uniform bounds on curvature and diameter. By the π_2-finiteness theorem (compare [PT]), if in the above example the diameter of the souls is normalized to one, then there is no such upper curvature bound for the souls.

Our starting point for gaining more control over the souls is the following:

Theorem 10.1.8. *([KPT]) There exists $D > 0$ such that in each dimension $n \geq 10$ there is an infinite sequence $(M_i^n)_{i\in\mathbb{N}}$ of pairwise homotopy equivalent but mutually non-homeomorphic closed simply-connected Riemannian n-manifolds with $0 \leq sec_{M_i^n} \leq 1$, positive Ricci curvature, and $diam_{M_i^n} \leq D$.*

The proof of this theorem proceeds along the following lines. Fix a rank 2 bundle ξ over $S^2 \times S^2 \times S^2$ and consider the sphere bundle P of $\xi \oplus \epsilon^{k-1}$, where $k \geq 3$. We then look at various circle bundles $S^1 \to M_i \to P$. A topological argument shows that with an appropriate choice of ξ, infinitely many such bundles have total spaces homotopy equivalent to $S^2 \times S^2 \times S^3 \times S^k$ but distinct first Pontrjagin classes, and thus are mutually non-homeomorphic. One can represent the M_i as $S^3 \times S^3 \times S^3 \times S^k/T_i^2$, where $T_i^2 \subset T^3$ acts freely and isometrically on $S^3 \times S^3 \times S^3 \times S^k$, and this easily implies that the M_i satisfy all the geometric properties claimed in the theorem.

Note that sequences of manifolds with these properties could theoretically also exist in dimension $n = 7, 8, 9$, but this is still unknown.

The above theorem deals with closed manifolds, which is not the principal topic in this chapter. However, by considering vector bundles over these closed manifolds, it is possible to find infinitely many examples with non-negative curvature and diffeomorphic total spaces. These bundles have the M_i as souls, thus providing the desired geometric control:

Theorem 10.1.9. *([KPT]) For any $k > 10$ the manifold $S^2 \times S^2 \times S^3 \times S^3 \times \mathbb{R}^k$ admits an infinite sequence of complete non-negatively curved metrics g_i with pairwise non-homeomorphic souls S_i such that*

$$0 \leq sec(M, g_i) \leq 1 \text{ and } diam(S_i) \leq D,$$

where D is a positive constant independent of k and i.

Examples with even better control over the codimensions of the souls were later produced by Belegradek, Kwasik and Schultz, improving on the above results from [Bel] and [KPT]:

Theorem 10.1.10. *([BKS]) For all $k \geq 3$ there are infinitely many complete Riemannian metrics with non-negative sectional curvature on $M = S^4 \times S^k \times \mathbb{R}^4$ whole souls are pairwise non-homeomorphic.*

§10.2 *Connectedness properties of $\mathcal{R}_{sec\geq 0}$ and $\mathcal{M}_{sec\geq 0}$*

To study connectedness properties of (moduli) spaces of non-negatively curved metrics, the following simple idea from [KPT] can be utilized. If one could show, possibly under some additional conditions, that (the equivalence classes of) complete non-negatively curved metrics with non-diffeomorphic souls are contained in different path components of $\mathcal{M}_{sec\geq 0}$, then the examples from the previous theorem could establish the existence of manifolds for which the spaces $\mathcal{R}_{sec\geq 0}$ and $\mathcal{M}_{sec\geq 0}$ possess an infinite number of path components.

We will see that when one works with the topology of C^k-convergence on compact subsets (see §1), where $0 \leq k \leq \infty$, then this is indeed true provided that any two complete metrics of non-negative sectional curvature have souls which intersect. Moreover, the latter condition can actually be forced by rather mild topological assumptions on the manifold M. For example this holds if M has a soul with nontrivial rational normal Euler class, or if M has a soul of codimension one (compare [KPT], [BKS]).

Theorem 10.2.1. *([KPT]) If (M, g_t), $t \in [0,1]$ is a continuous family of complete non-negatively curved metrics such that the normal bundle to the soul of (M, g_0) has nontrivial rational Euler class, then all the souls of (M, g_t) are diffeomorphic.*

Proof. Let S_t be the soul of (M, g_t). We claim that the family $(S_t, g_t|_{S_t})$ is continuous in Gromov-Hausdorff topology. Observe that since $S_t \hookrightarrow M$ is a homotopy equivalence, the rational Euler class of ν_{S_t} is nonzero for any t. Therefore it is sufficient to show that $S_t \overset{G-H}{\longrightarrow} S_0$ as $t \to 0$.

Let $\pi_t : M \to S_t$ be the Sharafutdinov retraction with respect to g_t, and let d_t be the inner metric on M induced by g_t.

Since $g_t \to g_0$ uniformly on compact sets, we clearly have that for any $x, y \in S_0$, $d_t(x,y) \leq d_0(x,y) + \epsilon_t$ where $\epsilon_t \to 0$ as $t \to 0$. Since π_t is distance nonincreasing, we see that $d_t(\pi_t(x), \pi_t(y)) \leq d_0(x,y) + \epsilon_t$ for any $x, y \in S_0$. Since $\pi_t : S_0 \to S_t$ is a homotopy equivalence of closed manifolds, as a degree one map it must be onto, and hence $\mathrm{diam} S_t \leq \mathrm{diam} S_0 + \epsilon_t$.

Now the assumption on the Euler class implies that $S_t \cap S_0 \neq \emptyset$ for any t. By the above all S_t have uniformly bounded diameters, and therefore they all must lie in some fixed closed ball $\bar{B}(p, D)$, where the ball is taken with respect to d_0. Again using the uniform convergence of g_t to g_0 on compact sets, we have that $d_0(x,y) \leq d_t(x,y) + \epsilon_t$ for any $x, y \in S_t$. Hence $d_0(\pi_0(x), \pi_0(y)) \leq d_t(x,y) + \epsilon_t$ for any $x, y \in S_t$. Combining this with the above we obtain:

$$d_0(\pi_0(\pi_t(x)), \pi_0(\pi_t(x))) \leq d_0(x,y) + 2\epsilon_t \text{ for any } x, y \in S_0.$$

This in turn implies that for some $\tilde{\epsilon}_t$ such that $\tilde{\epsilon}_t \to 0$ as $t \to 0$,

$$d_0(x,y) - 2\tilde{\epsilon}_t \leq d_0(\pi_0(\pi_t(x)), \pi_0(\pi_t(x))) \leq d_0(x,y) + 2\epsilon_t \text{ for any } x, y \in S_0.$$

Hence $\pi_0 \circ \pi_t : S_0 \to S_0$ is a $\max\{\epsilon_t, \epsilon_t\}$-Hausdorff approximation, and the same is true for $\pi_0 : (S_t, d_t) \to (S_0, d_0)$, which proves that $S_t \overset{G-H}{\longrightarrow} S_0$ as $t \to 0$.

Since S_t is a smooth manifold for any t and $\dim S_t = \dim S_0$, Yamaguchi's Stability theorem ([Yam]) then implies that S_t is diffeomorphic to S_0 for all t. □

By modifying the construction behind Theorem 10.1.9, it is possible to construct a manifold for which there are infinitely many non-negatively curved metrics for which the souls all have nontrivial rational Euler classes. This yields the following:

Theorem 10.2.2. *([KPT]) There exists an open manifold M^{22} which admits an infinite sequence of complete metrics g_i with pairwise non-homeomorphic souls S_i such that*

$$0 \leq sec(M, g_i) \leq 1 \text{ and } diam(S_i) \leq D,$$

and such that the equivalence classes of the metrics g_i all lie in different path components of the moduli space $\mathcal{M}_{sec\geq 0}(M)$. Moreover, for any closed non-negatively curved manifold (N, g), the product metrics $g_i + g$ all lie in different path components of $\mathcal{M}_{sec\geq 0}(M \times N)$.

Notice that the situation for positive *scalar* curvature is very different from this. Forming a product of M with any manifold (N, g) of positive scalar curvature, it is easy to construct a path of positive scalar curvature metrics between $g_i + g$ and $g_j + g$ for any i, j. Choose a smooth path of metrics $g_{i,j}(t)$, $t \in [0, 1]$, between g_i and g_j on M. There is no need for the metrics in this path to all have positive scalar curvature, but for reasons of smoothness we will assume that this path is constant for t close to both 0 and 1. By compactness of the interval, there is a sufficiently small $\lambda > 0$ so that $g_{i,j}(t) + \lambda^2 g$ has positive scalar curvature for all $t \in [0, 1]$. The desired path can then be formed by pre- and post-concatenating $g_{i,j}(t) + \lambda^2 g$ with smooth paths which shrink the scaling factor of g from 1 to λ and back again.

Let us now turn to the most significant later developments. Switching to the topology of uniform C^k-convergence and utilizing the smoothness of the Sharafutdinov retraction, in 2011 Belegradek, Kwasik and Schultz proved the following:

Theorem 10.2.3. *([BKS]) With respect to the topology of uniform C^k-convergence, $0 \leq k \leq \infty$, one has that*

(i) if two metrics are sufficiently close in $\mathcal{R}_{sec\geq 0}(M)$, their souls are ambiently isotopic in M;

(ii) the map associating to a metric $g \in \mathcal{R}_{sec\geq 0}(M)$ the diffeomorphism type of the pair $(M,$ soul of $g)$ is locally constant;

(iii) the diffeomorphism type of the pair $(M,$ soul of $g)$ is constant on connected components of $\mathcal{M}_{sec\geq 0}(M)$.

This result actually also holds verbatim for the topology of uniform C^k-convergence on compact subsets, *provided* any two metrics of non-negative sectional curvature on M have souls which intersect.

The following result can then be established using part (iii) of Theorem 10.2.3 and Theorem 10.1.10:

Corollary 10.2.4. *([BKS]) With respect to the topologies of uniform convergence, for all $k \geq 3$, $n \geq 4$, $\mathcal{M}_{sec\geq 0}(S^4 \times S^k \times \mathbb{R}^n)$ has infinitely many connected components.*

Finally we would like to mention that in the recent preprint [BFK] Belegradek, Farrell and Kapovitch showed that there are open manifolds for which the space of complete non-negatively curved metrics has higher homotopy groups of infinite order. There are currently no results of this kind for the moduli space $\mathcal{M}_{sec\geq 0}$.

§11 The Klingenberg-Sakai conjecture and the space of positively pinched metrics

As a sort of epilogue to the preceding parts of this book, we will discuss a conjecture of Klingenberg and Sakai in positive sectional curvature, concerning the existence of lower uniform bounds for the injectivity radius of δ-pinched metrics on manifolds in terms of the manifold and the pinching constant.

Uniform estimates for the injectivity radius of a Riemannian manifold have proved to be of great importance, for example, in the context of sphere and finiteness theorems. Here, however, we want to explain how this conjecture, when combined with results from [PRT] and [PT], can shed new light on the structure of the space of positively pinched and positively Ricci pinched metrics on a given manifold.

Conjecture 11.1. *(The Klingenberg-Sakai conjecture [KlSa2]) Let M be a smooth closed simply-connected manifold and let $0 < \delta \le 1$. Then there exists $i_0 = i_0(M, \delta) > 0$ such that the injectivity radius i_g of any δ-pinched metric g on M, i.e., any Riemannian metric with sectional curvature $\delta \le sec_g \le 1$, is bounded from below by i_0.*

In one of his problem lists, S.-T. Yau has also posed the following more general and closely related question:

Question 11.2. *([Yau]) Let M be a smooth closed simply-connected manifold and let $0 < \delta \le 1$. Does there exist a positive number i_0 depending only on δ and the homotopy type of M, such that the injectivity radius i_g of any δ-pinched metric g on M is bounded from below by i_0?*

In addition, both of these problems also make sense for *positive Ricci pinching*, i.e., for metrics g on m-dimensional closed smooth manifolds satisfying the less restrictive curvature bounds $sec_g < 1$ and $Ric_g \ge (m-1)\delta > 0$.

The Klingenberg-Sakai conjecture is presently known to be true in the following principal cases:

(i) The dimension m is even and $\delta > 0$ is arbitrary (Klingenberg [Kl1] 1959). (The case $m = 2$ was already solved by Pogorelov [Po] in 1946.)

(ii) The dimension m is odd and for $\delta \ge 1/4 - \epsilon$, $\epsilon \approx 10^{-6}$ (Abresch-Meyer [AM] 1994). (For odd dimensions m, the case $\delta > 1/4$ was solved by Klingenberg [Kl2] in 1961, and the case $\delta > 1/4$ was treated independently by Cheeger-Gromoll [CG2] and Klingenberg-Sakai [KlSa1] in 1980.)

(iii) The manifold has finite second homotopy group, the dimension m is arbitrary, $sec_g \le 1$ and $Ric_g \ge (m-1)\delta > 0$, where $\delta > 0$ is arbitrary (Petrunin-Tuschmann [PT] 1999). (For $m = 3$ this was already shown by Burago-Toponogov [BT] in 1973 - see also Sakai's work [Sa]. Moreover, for general dimensions m, the special case of sectional curvature pinching $0 < \delta \le sec_g \le 1$ was solved independently by Fang-Rong [FR] and Petrunin-Tuschmann [PT] in 1999.)

W. Tuschmann, D.J. Wraith, *Moduli Spaces of Riemannian Metrics*,
Oberwolfach Seminars 46, DOI 10.1007/978-3-0348-0948-1_11

In all of the results mentioned in (i) and (ii) above, i_0 can in fact be chosen to be independent of the manifold M and the precise value of δ, namely, $i_0 = \pi$. On the other hand, in (iii) i_0 does not depend on the topology of M, but in general it *does* depend on both δ and m. In fact, the Berger spheres demonstrate that in general (including dimension three), i_0 will depend on δ. (See [AM].)

The existence of uniformly pinched collapsing sequences among the Aloff-Wallach, Eschenburg and Bazaikin spaces (see [AW], [Es], [Ba]) - which all have second homotopy groups of infinite order - show that for small positive δ (in fact $\delta < 1/37$ will work, see [Pü]) there is no chance for the conjecture to hold if one does not fix the topology of M.

Under the more general conditions of positive Ricci pinching $sec_g \leq 1$ and $Ric_g \geq (m-1)\delta > 0$, for the existence of uniform positive lower bounds on the injectivity radius it is actually necessary that the second homotopy groups of the manifolds be finite, even if one fixes their topological type. Namely, in [PT] it is shown that there is a sequence of metrics g_n on $S^2 \times S^3$ which satisfy the bounds $sec_{g_n} \leq 1$ and $Ric_g \geq 4\delta > 0$, but for which the spaces $(S^2 \times S^3, g_n)$ collapse to $S^2 \times S^2$ as $n \to \infty$.

For $m > 3$ odd and $\delta > 0$ arbitrary, without additional restrictions as in (iii) above, the Klingenberg-Sakai conjecture is completely open. To describe further results, note first that one can reformulate the Klingenberg-Sakai conjecture in terms of Gromov-Hausdorff convergence as follows:

Conjecture 11.3. *Suppose that a compact simply-connected manifold M admits a sequence of metrics (g_n) with sectional curvature $\lambda \leq K_{g_n} \leq \Lambda$, such that as $n \to \infty$, the sequence of metric spaces (M, g_n) Gromov-Hausdorff converges to a compact metric space X of lower dimension (i.e. $dimX < dimM$). Then $\lambda \leq 0$ (that is, these metrics cannot be uniformly positively pinched).*

Definition 11.4. *A sequence of metric spaces M_i is called stable if there is a topological space M and a sequence of metrics d_i on M such that (M, d_i) is isometric to M_i and the metrics d_i converge as functions on $M \times M$ to a continuous pseudometric.*

Theorem 11.5. *(Stable Collapse [PRT]) Suppose that a compact (topological) manifold M admits a stable sequence of Riemannian metrics (g_n) with sectional curvatures $\lambda \leq K_{g_n} \leq \Lambda$, such that as $n \to \infty$ the metric spaces (M, g_n) Gromov-Hausdorff converge to a compact metric space X of lower dimension. Then $\lambda \leq 0$ (that is, these metrics cannot be uniformly positively pinched).*

Here is an equivalent, but 'convergence-free' version of this result:

Theorem 11.6. *(Bounded Version of the Klingenberg-Sakai Conjecture [PRT]) Let M be a closed (topological) manifold, suppose that d_0 is a metric on M, and let $0 < \delta \leq 1$. Then there exists $i_0 = i_0(M, d_0, \delta) > 0$ such that the injectivity radius i_g of any δ-pinched d_0-bounded metric g on M, i.e., any Riemannian metric with sectional curvature $\delta \leq K_g \leq 1$ and $dist_g(x,y) \leq d_0(x,y)$, is bounded from below by i_0.*

Recall that the Berger spheres constitute an example of a collapse of S^{2m+1} to $\mathbb{C}P^m$ by a continuous one parameter family of Riemannian metrics with positive curvature $0 < K \leq 1$. The next result shows in particular that under the assumption of *positive pinching*,

$0 < \delta \leq K_g \leq 1$, such phenomena cannot occur. Thus we have a *continuous version of the Klingenberg-Sakai conjecture*:

Theorem 11.7. *(Continuous Collapse [PRT]) Suppose that a compact manifold M admits a continuous one parameter family $(g_t)_{0<t\leq 1}$ of Riemannian metrics with sectional curvature $\lambda \leq K_{g_t} \leq \Lambda$, such that as $t \to 0$ the family of metric spaces (M, g_t) Gromov-Hausdorff converges to a compact metric space X of lower dimension. Then $\lambda \leq 0$ (that is, these metrics cannot be uniformly positively pinched).*

This last result in particular implies the following. Suppose that the Klingenberg-Sakai conjecture is false, so that there exists an m-dimensional manifold M which, for some $\delta > 0$, admits a collapsing sequence $\{g_n\}$ of δ-pinched metrics. We may then assume that the sequence of metric spaces (M, g_n) Gromov-Hausdorff converges to a compact metric space X of lower dimension. Then there exists an $\epsilon = \epsilon(M, \delta, X) > 0$, such that the intersection of the space of all δ-pinched metrics g on M with the Gromov-Hausdorff ϵ-neighborhood of X has an infinite number of connected components. Moreover, for each of these components its infimum distance to X is positive (see the picture below).

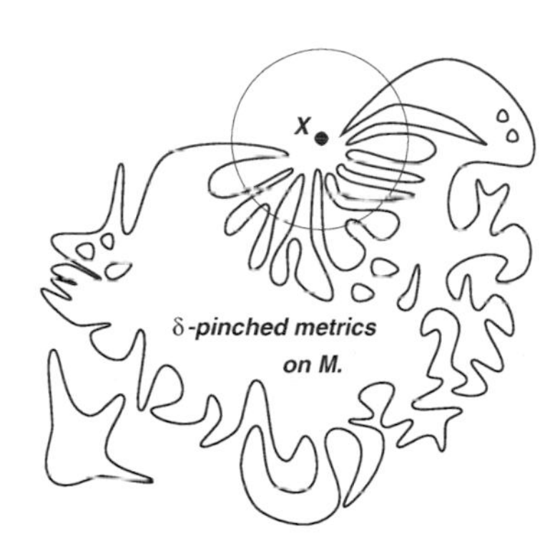

By results in [PT], the Stable Collapse Theorem (11.5), the Bounded Version of the Klingenberg-Sakai Conjecture (11.6), and the Continuous Collapse Theorem (11.7) all continue to hold when the uniform positive pinching of sectional curvature $0 < \delta \leq sec_g \leq 1$ is replaced by the positive Ricci pinching conditions $sec_g \leq 1$ and $Ric_g \geq (m-1)\delta > 0$. In particular, one is again in the situation illustrated by the picture above.

Appendix A
K-Theory and the α-invariant

In this appendix we outline some basics of K-theory and related issues which appear in the text. These related issues are the Atiyah-Bott-Shapiro construction, the index of a Cl_k-linear Dirac operator, the definition of the α-invariant, and the existence of exotic spheres with non-zero α-invariant. For more details see, for example the books [A] or [Hu] for K-theory, [ABS] or [LM; I.9] for the Atiyah-Bott-Shapiro construction, and [M2] or [LM] (especially page 100 and §2.7) for more about the α-invariant.

Throughout this appendix we will consider bundles over compact base spaces only.

A.1 *K-Theory.*

Denote by $Vect(X)$ the set of isomorphism classes of complex vector bundles over the (compact) space X. The complex K-theory group $K(X)$ can be formed from $Vect(X)$ by the 'Grothendieck group' construction as follows. Let $F(X)$ be the free-abelian group generated by the elements of $Vect(X)$, so $F(X)$ involves formal sums and differences of isomorphism classes. Now form the quotient of $F(X)$ by the subgroup $E(X)$ generated by elements of the form $V + W - V \oplus W$, where V and W are isomorphism classes over X. Thus in the quotient group '+' corresponds to the direct sum operation. This quotient group is $K(X)$.

Given an isomorphism class of bundles V, we will write $[V]$ for the corresponding class in $K(X)$. Notice that we can represent any element of $K(X)$ by a difference of classes $[V] - [W]$. Consider the equation $[V] - [W] = 0$. By definition this means that $V - W \in E(X)$, i.e. there are isomorphism classes of bundles P_i, Q_i, R_j, S_j such that

$$V - W = \sum_i (P_i + Q_i - P_i \oplus Q_i) - \sum_j (R_j + S_j - R_j \oplus S_j)$$

in $F(X)$. (Here the right-hand side represents a generic element of $E(X)$.) Rearranging we have

$$V \oplus \sum_i P_i \oplus Q_i + \sum_j R_j + \sum_j S_j = W + \sum_i P_i + \sum_i Q_i + \sum_j R_j \oplus S_j$$

in $F(X)$. Using the same symbols to denote representatives of the isomorphism classes of bundles, this statement is equivalent to $V \oplus E \cong W \oplus E$, where E is the bundle $\sum_{i,j} P_i \oplus Q_i \oplus R_j \oplus S_j$. Now it is well-known (see for example [LM; page 60]) that every bundle has a 'complementary bundle' such that the direct sum of the bundle with its complement is trivial. In particular, there is a bundle $E^\perp$ such that $E \oplus E^\perp \cong n$, where n denotes the trivial bundle of dimension n over X, i.e. $X \times \mathbb{C}^n$. Thus we deduce that $[V] = [W]$ in $K(X)$ if and only if $V \oplus n \cong W \oplus n$ for some trivial bundle n, that is, if and only if V and W are *stably isomorphic.*

W. Tuschmann, D.J. Wraith, *Moduli Spaces of Riemannian Metrics*,
Oberwolfach Seminars 46, DOI 10.1007/978-3-0348-0948-1

More generally, given an element $[V]-[W] \in K(X)$, we have

$$[V]-[W]=[V]+[V^{\perp}]-([W]+[V^{\perp}])=[V \oplus V^{\perp}]-[W \oplus V^{\perp}]=[n]-[W \oplus V^{\perp}]$$

for some trivial bundle n. Thus any element of $K(X)$ can be written (non-uniquely) in the form $[n]-[E]$ for a trivial bundle n and a bundle E.

As a simple example, consider $K(pt)$. As there is only the trivial bundle in each dimension over a point, it is clear that we have an isomorphism $K(pt) \cong \mathbb{Z}$, where the element $[n]-[m]$ is mapped to $n-m \in \mathbb{Z}$.

Notice that we have not used the fact that the fibres in our bundles are *complex* vector spaces. Thus if we wish to consider real vector bundles over X, exactly the same construction goes through. However the resulting K-groups are distinguished from the complex case by using the notation $KO(X)$.

Given a continuous map $f : X \to Y$, it is straightforward to see that the pull-back construction for bundles induces a homomorphism of K-groups $f^* : K(Y) \to K(X)$. With a little further work it can be show that the homomorphism f^* only depends on f up to homotopy. As an immediate corollary we see that if $X \simeq Y$ then $K(X) \cong K(Y)$.

A.2 *The difference bundle construction.*

We begin by defining the *reduced K-groups* for a based space. If x_0 denotes the basepoint of X, we define $\tilde{K}(X)$ to be the kernel of the homomorphism $K(X) \to K(x_0)$ induced by the inclusion of x_0 into X. If X is a connected space we could equally define $\tilde{K}(X)$ to be the subgroup of $K(X)$ consisting of elements which can be represented in the form $[n]-[V]$, where $\dim V = n$. For example it is a triviality that $\tilde{K}(pt)=0$.

If A is a closed subset of X, define the relative K-group $K(X,A) := \tilde{K}(X/A)$. It can be shown (see [A; page 69]) that there is an exact sequence

$$K(X,A) \to K(X) \to K(A).$$

(In fact such a sequence fits into a long exact sequence described in the section below, by virtue of the fact that K-theory is a generalized cohomology theory.) Note that the map $K(X,A) \to K(X)$ is not in general injective.

Let us consider the relationship between elements of $K(X,A)$ and those of $K(X)$. If a bundle V over X happens to be trivial over $A \subset X$, then intuitively we should be able to 'collapse' the bundle over A to get a bundle with base X/A. However there is a problem: this collapsing involves identifying the fibres over A, but how do we do this? We need to use an explicit trivialization $T : V|_A \to A \times \mathbb{C}^n$, but in general there will be many of these, with any two trivializations differing by a map $Y \to GL(n,\mathbb{C})$ for some n. Such maps could be homotopically non-trivial. Thus we can obtain elements of $K(X,A)$ starting from bundles over X equipped with a trivialization over A. Conversely, given a bundle over X/A we can pull back to a bundle over X using the quotient map $X \to X/A$. Any fibre over A in such a pull-back bundle can be canonically identified with an element in the fibre over $A/A \in X/A$, and thus the pull-back comes with a trivialization over A.

The *difference bundle construction* asserts that we can represent any element of $K(X, A)$ by an equivalence class $[V_0, V_1; \sigma]$ of triples $(V_0, V_1; \sigma)$, where the V_i are vector bundles over X, and σ is a vector bundle isomorphism between $V_0|_A$ and $V_1|_A$. The role of σ here is to deal with the trivialization issues discussed above. The equivalence relation is given by $(V_0, V_1; \sigma) \sim (W_0, W_1; \tau)$ if and only if there are bundles P and Q and isomorphisms $\iota_0 : V_0 \oplus P \cong W_0 \oplus Q$ and $\iota_1 : V_1 \oplus P \cong W_1 \oplus Q$ such that $(\tau \oplus id_Q) \circ \iota_0 = \iota_1 \circ (\sigma \oplus id_P)$. In particular this means that $[V_0, V_1; \sigma] = [V_0 \oplus W, V_1 \oplus W; \sigma \oplus \mathrm{id}_W]$ for any vector bundle W over X. It also means that if we have bundle isomorphisms $\phi_i : V_i \cong W_i$, for $i = 0, 1$, then $[V_0, V_1; \sigma] = [W_0, W_1; \phi_1 \circ \sigma \circ \phi_0^{-1}]$. In other words, we can replace any bundle in a triple by an isomorphic one provided we adjust the isomorphism over A appropriately.

Denote by $L(X, A)$ the set of all equivalence classes of triples. It is clear that $L(X, A)$ is a semigroup under the obvious direct sum operation. With a little more work it can be shown that $L(X, A)$ is an abelian group. The identity element is the class of triples $(V_0, V_1; \sigma)$ where σ can be extended to a vector bundle isomorphism over all of X. The inverse of an element $[V_0, V_1; \sigma]$ is given by $[V_1, V_0; \sigma^{-1}]$, see for example [LuMi; page 139].

There is an isomorphism of abelian groups $L(X, A) \to K(X, A)$. A standard argument establishes this isomorphism, and this can be found in [ABS], [A; §2.6] or [LM; I.9]. However we give an equivalent, but slightly different description which has the benefit of being more explicit.

Given a triple $[V_0, V_1; \sigma]$, consider a complementary bundle $V_1^{\perp}$ to V_1. Denoting the identity map of $V_1^{\perp}$ by 1, we have $[V_0, V_1; \sigma] = [V_0 \oplus V_1^{\perp}, V_1 \oplus V_1^{\perp}; \sigma \oplus 1]$. There is a bundle isomorphism $\iota : V_1 \oplus V_1^{\perp} \cong X \times \mathbb{C}^n$, which then gives

$$[V_0, V_1; \sigma] = [V_0 \oplus V_1^{\perp}, X \times \mathbb{C}^n; \iota \circ (\sigma \oplus 1)].$$

Now we use $\iota \circ (\sigma \oplus 1)$ to collapse the bundle $V_0 \oplus V_1^{\perp}$ over A. Thus we obtain a bundle E together with a trivial bundle $[n]$ over X/A, and hence we can define an element $[E] - [n] \in K(X, A)$. It can be shown that this construction is well-defined, i.e. is independent of the various choices made, and with a little more work one can prove that this correspondence gives an isomorphism between $L(X, A)$ and $K(X, A)$. For further details see [Hu; Theorem 5.1].

The relevance of the difference bundle construction for our purposes arises from the so-called 'Atiyah-Bott-Shapiro construction', which we will now explain.

A.3 *The Atiyah-Bott-Shapiro construction.*

Consider the set of isomorphism classes of $\mathbb{Z}_2$-graded modules for the Clifford algebra Cl_n. Notice that this has a natural semigroup structure under direct sum. In a similar manner to that in which we defined the groups $K(X)$, consider the Grothedieck group generated by this semigroup. Call this group $\mathcal{M}_n^c$ in the complex setting, and $\mathcal{M}_n$ in the real case.

Given a $\mathbb{Z}_2$-graded module $W = W^0 \oplus W^1$ for Cl_n, we construct a triple $[E_0, E_1; \mu] \in K(D^n, S^{n-1})$ as described in A.2 above, by setting $E_i = D^n \times W^i$, $i = 0, 1$, and where $\mu : E_0|_{S^{n-1}} \to E_1|_{S^{n-1}}$ is given by Clifford multiplication of the W^0 fibres of E_0 by the corresponding element in ∂D^n : $\mu(x, w) = (x, x \cdot w)$. In fact, this formula makes μ an isomorphism over all of $D^n \setminus \{0\}$, but clearly we do *not* have an isomorphism over the fibres at 0. However, if the Cl_n-action on W is actually the restriction of a Cl_{n+1} action on W, then we *can* extend μ to an isomorphism over all of D^n using the action of the 'extra dimension' e_{n+1}. We set $\mu(x, w) = (x, (x + e_{n+1}\sqrt{1 - |x|^2}) \cdot w)$. Thus in the complex case (and similarly in the real), our map $\mathcal{M}_n^c \to K(D^n, S^{n-1})$ descends to a map

$$\mathcal{M}_n^c / i^* \mathcal{M}_{n+1}^c \to K(D^n, S^{n-1}),$$

where $i : \mathcal{M}_n^c \to \mathcal{M}_{n+1}^c$ is the inclusion map. It turns out that this map is an isomorphism of abelian groups.

In fact we can say more, but first a little topology. Given two based spaces X and Y, we can form the 'smash product' $X \wedge Y := X \times Y / X \vee Y$, where we use basepoints x_0 and y_0 to construct $X \vee Y$:

$$X \vee Y := (X \times \{x_0\}) \cup (\{y_0\} \times Y) \subset X \times Y.$$

It can be shown that for spheres we have a homeomorphism $S^n \wedge S^m \cong S^{n+m}$, and for any space X it is a trivial consequence of the definition that $X \wedge \{pt\} = \{pt\}$ and $X \wedge S^0 \cong X$. After choosing a basepoint on the sphere S^n, we define the 'reduced suspension' of X, ΣX to be $S^1 \wedge X$, and the k-fold suspension $\Sigma^k X := S^k \wedge X$.

Let $\tilde{K}^{-k}(X) := \tilde{K}(\Sigma^k X)$ and $K^{-k}(X, A) := \tilde{K}(\Sigma^k(X/A))$. So for example

$$K(D^n, S^{n-1}) = \tilde{K}(D^n / \partial D^n) = \tilde{K}(S^n) = K^{-n}(S^0),$$

since $S^k \wedge S^0 \cong S^k$. It can be shown that there is a long exact sequence

$$\ldots \to \tilde{K}^{-k-1}(A) \to K^{-k}(X, A) \to \tilde{K}^{-k}(X) \to \tilde{K}^{-k}(A) \to K^{-k+1}(X, A) \to \ldots$$

The sequence terminates with $\tilde{K}^0$ terms, where $\tilde{K}^0$ is identified with $\tilde{K}$.

In the situation where X is not equipped with a basepoint, we consider the union of X and a disjoint basepoint, $X^+ := X \cup \{pt\}$. It is easy to see that $K(X) \cong \tilde{K}(X^+)$. We further define

$$K^{-k}(X) = K^{-k}(X, \emptyset) := \tilde{K}(\Sigma^k(X^+)).$$

In the special case where X is itself a point, we have $K^{-k}(pt) = \tilde{K}(S^k \wedge S^0) \cong \tilde{K}(S^k)$.

It can be shown that the tensor product operation on bundles induces a map

$$K^{-i}(X) \otimes K^{-j}(Y) \to K^{-(i+j)}(X \times Y).$$

As a consequence of this we see that $K^{-*}(pt)$ has the structure of a graded ring, and moreover for any space X, $K^{-*}(X)$ is a graded module over $K^{-*}(pt)$. The Bott Periodicity Theorem (see for example [LM; I.9]) sheds light onto these ring and module structures. A simplified version of this theorem is as follows:

Theorem A.3.1. *(Bott Periodicity). For any compact space X we have a module isomorphism*

$$K^{-k}(X) \cong K^{-(k+2)}(X)$$

in the complex case, and

$$KO^{-k}(X) \cong KO^{-(k+8)}(X)$$

in the real case. Ignoring the ring structure, the groups $K^{-*}(pt)$ and $KO^{-*}(pt)$ are as follows: $K^{-2k}(pt) \cong \mathbb{Z}$ and $K^{-(2k+1)}(pt) = 0$ for all $k \geq 0$. In the real case we have $KO^{-k}(pt)$ is isomorphic to $\mathbb{Z}$ if $k \equiv 0$ or 4 mod 8, to $\mathbb{Z}_2$ if $k \equiv 1$ or 2 mod 8, and is zero otherwise.

Recall that periodicity phenomena of order two (respectively eight) appear in the study of complex (respectively real) Clifford algebras and modules. This is not a coincidence, and is explained by the following theorem of Atiyah, Bott and Shapiro:

Theorem A.3.2. *([ABS]) There are graded ring isomorphisms*

$$\mathcal{M}^c_*/i^*\mathcal{M}^c_{*+1} \cong K^{-*}(pt);$$

$$\mathcal{M}_*/i^*\mathcal{M}_{*+1} \cong KO^{-*}(pt).$$

A.4 *The index of Cl_k-linear Dirac operators.*

In §3.3 we mention the existence of a so-called Cl_k-linear Dirac operator. This is a Dirac operator on a spinor bundle S over a manifold X, with the property that Cl_k acts from the right on S by fibre-preserving automorphisms, and this action commutes with multiplication by elements of $Cl(X)$. Moreover we can consider such spinor bundles which are $\mathbb{Z}_2$-graded, which means that S is simultaneously $\mathbb{Z}_2$-graded with respect to both left $Cl(X)$ and right Cl_k-actions.

Given a spin manifold M^k, there is a canonical bundle over M with this property, and hence a canonical Cl_k-linear Dirac operator. The bundle is $P_{Spin} \times_l Cl_k$, where P_{Spin} is the principal Spin_k-bundle associated to TM, and l denotes the action of Spin_k on Cl_k by left multiplication.

We noted in §3.3 that a Cl_k-linear Dirac operator has a notion of index which takes values in $KO^{-k}(pt)$. We are now in a position to both give and explain this definition.

If D denotes a Cl_k-linear Dirac operator on a $\mathbb{Z}_2$-graded spinor bundle, we have as usual $D = D^0 \oplus D^1$. Observe that the kernel $\ker D$ is a $\mathbb{Z}_2$-graded (right) Cl_k-module, and thus by A.3 defines a class $[\ker D] \in \mathcal{M}_k/i^*\mathcal{M}_{k+1}$. Using the Atiyah-Bott-Shapiro isomorphism we can then define the Cl_k-linear index $\mathrm{ind}_k D$ to be the image of $[\ker D]$ in $KO^{-k}(pt)$.

Despite the fact that this index looks somewhat different from the other indexes considered in this book, it is in fact a generalization of the usual index as we will now explain. We claim that if we identify $KO^0(pt)$ in the usual way with the integers, then the Cl_0-index agrees with the normal index. To see this, recall that $Cl_0 = \mathbb{R}$, and $Cl_1 = \mathbb{C}$. Thus a Cl_0-module is just a real vector space, and a $\mathbb{Z}_2$-graded Cl_0-module is therefore a pair of real vector spaces $V^0 \oplus V^1$. Now this sum belongs to the image of $i^*\mathcal{M}_1$ if and only if $V^0 \oplus V^1$ has the structure of a complex vector space, where $i \in \mathbb{C}$ acts by swapping the summands. In particular this means that V^0 and V^1 must have equal dimensions. This occurs in particular when $V^0 = V^1 = V$, since $V \oplus V \cong V \otimes \mathbb{C}$. As a consequence we see that $[V \oplus 0] = -[0 \oplus V] \in \mathcal{M}_k / i^*\mathcal{M}_{k+1}$. With this in mind we compute:

$$\begin{aligned} \text{ind}_0 D &= [\ker D] \\ &= [\ker D^0 \oplus \ker D^1] \\ &= [\ker D^0 \oplus 0] + [0 \oplus \ker D^1] \\ &= [\ker D^0 \oplus 0] - [\ker D^1 \oplus 0]. \end{aligned}$$

As the composition of the Atiyah-Bott-Shapiro isomorphism with the canonical isomorphism $KO^0(pt) \cong \mathbb{Z}$ just computes (differences of) dimensions, we see that the Cl_0 index is just $\dim\ker D^0 - \dim\ker D^1$, which agrees with the usual index $\dim\ker D^0 - \dim\text{coker} D^0$ using the fact that D^0 and D^1 are adjoints of each other.

As noted in §3.3, if we write the canonical Cl_k-linear Dirac operator as $D = D^0 \oplus D^1$, we obtain a restricted operator D^0 called the Atiyah-Singer Cl_k-linear Dirac operator. There is an Atiyah-Singer index theorem for this (see [LM;III.16.6]):

Theorem A.4.1. *For a compact spin manifold M^k, the index of the Atiyah-Singer Cl_k-linear Dirac operator is given by*

$$ind_k(M) = \alpha(M),$$

where $\alpha(M)$ is the α-invariant of M.

The α-invariant (see for example §4.1.2 and §3.3) can be defined topologically in a way we will now describe.

A.5 *The α-invariant.*

Given a vector bundle E, we can consider the corresponding disc and sphere bundles $D(E)$ and $S(E)$ (as determined by an arbitrary background Riemannian metric). The Thom space of E is defined to be the quotient space $D(E)/S(E)$.

For fixed n and large k there is an isomorphism $\Omega_n^{spin} \cong \pi_{n+8k} M\text{Spin}(8k)$, between the n-dimensional spin bordism group and a homotopy group of a certain topological space $M\text{Spin}(8k)$. This space is the Thom space of the universal $8k$-plane bundle over $B\text{Spin}(8k)$, which is the $\mathbb{R}^{8k}$-bundle associated to the universal $\text{Spin}(8k)$-bundle $E\text{Spin}(8k)$ using the canonical $\text{Spin}(8k)$-action on $\mathbb{R}^{8k}$ (i.e. via the double covering map $\text{Spin}(8k) \to SO(8k)$). The map between spin bordism and the homotopy groups of this Thom space is given by the Pontrjagin-Thom construction, which proceeds as follows. Embed M into a high dimensional Euclidean space, which should be viewed as a sphere less a point. Consider the normal disc bundle of the image of the embedding, and collapse out the complement. This gives a map from the sphere to the Thom space of the (stable) normal bundle of the image.

Now consider the irreducible real spinor bundle S associated to the universal $8k$-plane bundle E_{8k}. Split this into $S^{\pm}$ and pull back to the disc bundle of E_{8k} using the projection map to $B\mathrm{Spin}(8k)$. Thus we get two bundles over $D(E_{8k})$ for which the restrictions to the boundary are isomorphic via Clifford multiplication by the disc vector at each boundary point. This gives a difference element in KO of the Thom space. So far this is all canonical.

Next pull back this KO-theory element via a map $f \in \pi_{n+8k}M\mathrm{Spin}(8k)$ representing a given spin bordism class in Ω_n^{spin}. This gives an element in $KO(S^{n+8k}) \cong KO(S^n) = KO^{-n}(pt)$. This is the α-invariant of the spin bordism class.

A.6 *Construction of exotic spheres with $\alpha \neq 0$.*

Recall that such spheres, discovered by Hitchin [Hi], cannot admit a metric of positive scalar curvature (see §3.4).

We start from the fact that there is a 'bad' spin structure on S^1 which has non-zero α-invariant. Here, the bad spin structure is two copies of S^1, and it is bad because it does not extend over the disc. The alpha invariant for S^1 equipped with this spin structure is $1 \in \mathbb{Z}_2$. There are two spin structures on S^1. The other structure (the 'good' spin structure) *does* extend over the disc, making S^1 equipped with this spin structure a spin boundary. The α-invariant for this spin manifold must therefore vanish.

Now form the product of S^1 with a manifold M^{4n} constructed as follows. Consider the plumbing (see §4.2.1) of eight copies of the tangent disc bundle to S^{2n} according to the E_8 graph. Form the boundary connected sum of k copies of this manifold, where k is the order of the group bP_{4n} of homotopy spheres which bound a parallelisable manifold. The boundary of the resulting manifold is a sphere, and to this we can glue a disc D^{4n} to obtain a closed manifold. This is the manifold M^{4n}. (See also §6.2.2.) It has a unique spin structure and non-vanishing α invariant, as its $\hat{A}$ genus can be shown to be non-zero. Now we use surgery to kill the homotopy groups of $M \times S^1$ while preserving the spin structure. (The non-trivial homotopy groups are π_1 and π_{2n}.) The fact that surgery can be used here is due to the fact that M is parallelisable over the complement of a point (which can be taken to be the centre of the disc D^{4n}). In particular all the spheres on which we wish to do surgery have trivial normal bundles. The surgery process results in a homotopy sphere which is spin bordant to the original product, and thus has non-zero α-invariant, as required. It also follows that such spheres cannot be spin boundaries.

Appendix B

An overview of the Atiyah-Patodi-Singer index theorem

The Atiyah-Patodi-Singer index theorem [APS1] plays a crucial role in the development of the Kreck-Stolz s-invariant as discussed in §5. In this appendix we provide some more details of the ideas behind this index theorem. The main approach to the proof involves the concept of 'heat kernels', and we begin with a discussion of these.

B.1 *Heat kernels*

The heat equation in $\mathbb{R}^n$ takes the form $\partial f/\partial t = \Delta u$ where Δ is the Laplacian on $\mathbb{R}^n$ and $f = f(x,t)$. This can be generalized in the following way. For a Riemannian bundle E over a compact manifold X, let $D : \Gamma(E) \to \Gamma(E)$ be a non-negative (i.e. with only non-negative eigenvalues) self-adjoint elliptic differential operator of order m. Then there is a corresponding 'heat equation'

$$\frac{\partial f}{\partial t} + Df = 0.$$

The *heat operator* $e^{-tD} : \Gamma(E) \to \Gamma(E)$ is defined by

$$e^{-tD}\tilde{f}(x) := \int_X K(t,x,y)\tilde{f}(y)\,dy$$

where $\tilde{f} \in \Gamma(E)$, $x, y \in X$, $t > 0$, and where $K(t,x,y) : E_y \to E_x$ is a certain linear, smoothly varying function. $K(t,x,y)$ is the so-called *heat kernel*, and is the fundamental solution to the heat equation. The effect of the heat operator is to transform sections $\tilde{f}(x)$ of E into solutions $f(x,t)$ for the heat equation. If we think of the section $\tilde{f}$ as the initial condition $f(x,0)$, then the solution to the corresponding Cauchy problem (i.e. with solution value specified on a hypersurface - in this case $t = 0$) is obtained by convolution with K. Note that this solution is unique among solutions which grow slower than e^{t^2} in t.

The heat kernel has an expansion:

$$K(t,x,y) = \sum_{k=1}^{\infty} e^{-\lambda_k t}\langle f_k(y), -\rangle f_k(x)$$

where $\{f_k\}$ is a complete orthonormal basis for $L^2(E)$, consisting of eigenfunctions of D with corresponding eigenvalues λ_k (where $0 \leq \lambda_1 \leq \lambda_2 \leq \lambda_3 \leq ....$). There is also an asymptotic expansion (see [Se]) for an n-dimensional compact manifold as $t \to 0^+$:

$$K(t,x,y) = \frac{1}{(4\pi t)^{n/m}}e^{-d(x,y)^2/4t}\Big(c_0(x,y) + c_1(x,y)t^{1/m} + ... + c_k(x,y)t^{k/m} + O(t^{(k+1)/m})\Big)$$

W. Tuschmann, D.J. Wraith, *Moduli Spaces of Riemannian Metrics*,
Oberwolfach Seminars 46, DOI 10.1007/978-3-0348-0948-1

where $d(x,y)$ is the Riemannian distance from x to y, each c_j is smooth on $M \times M$ and $c_j(x,y) \in \mathrm{Hom}(E_y, E_x)$.

Of particular importance is the trace of the heat kernel:

$$\begin{aligned}\mathrm{tr}(e^{-tD}) &= \int_X \mathrm{trace}_x K(t,x,x)\,dx \\ &= \sum_{k=1}^{\infty} e^{-\lambda_k t}.\end{aligned}$$

Using the asymptotic expansion for compact manifolds as $t \to 0^+$ we have

$$\begin{aligned}\mathrm{tr}(e^{-tD}) =& \frac{1}{(4\pi t)^{n/m}}\Big(\int_X \mathrm{tr}c_0(x,x)\,dx + t^{1/m}\int_X \mathrm{tr}c_1(x,x)\,dx + \dots \\ &\dots + t^{k/m}\int_X \mathrm{tr}c_k(x,x)\,dx + O(t^{(k+1)/m})\Big).\end{aligned}$$

Notice that the term $k = n$ in the above formula is a constant term (i.e. independent of t). Comparing this with the expression for $\mathrm{tr}(e^{-tD})$ as a sum of exponentials, since $\mathrm{dimker}D = \sum e^{-t\lambda_i}$ where i ranges over all the zero eigenvalues of D we see immediately that

$$\dim\ \mathrm{ker}D = \int_X \mathrm{tr}c_n(x,x)\,dx.$$

If we now let D denote any elliptic differential operator, the operators D^*D and DD^* are then non-negative, elliptic, and also self-adjoint. Consequently D^*D and DD^* have the same set of non-zero eigenvalues. Notice also that $\mathrm{ker}D = \mathrm{ker}D^*D$: we have $(D^*Dx, x) = (Dx, Dx) = |Dx|^2$ so $D^*Dx = 0$ if and only if $Dx = 0$. Similarly we have $\mathrm{ker}D^* = \mathrm{ker}DD^*$. Thus

$$\mathrm{ind}D = \mathrm{dimker}D - \mathrm{dimker}D^* = \mathrm{dimker}D^*D - \mathrm{dimker}DD^*.$$

Since the non-zero eigenvalues of D^*D and DD^* are the same we obtain the basic formula

$$\mathrm{ind}D = \mathrm{tr}e^{-tD^*D} - \mathrm{tr}e^{-tDD^*}.$$

Of course we can interpret the right hand side in terms of the asymptotic expansion above. As the index is independent of t, evaluating the right-hand side for any t will yield the index. Equivalently we see that the index is also given by the difference of the constant terms in the asymptotic expansion. Evaluating this is essentially the content of the Atiyah-Singer index theorem.

B.2 *Atiyah-Patodi-Singer: the heat kernel approach*

We want to compute the index of a linear first-order elliptic differential operator D acting on a compact manifold X with boundary Y. This operator takes the form

$$D : C^\infty(X,E) \to C^\infty(X,F),$$

where E and F are vector bundles over X, and where we have written, for example, $C^\infty(X, E)$ instead of $\Gamma(E)$ in order to be clear about base spaces. We will assume that E and F are equipped with smoothly varying inner products in the fibres, which are independent of the normal parameter near the boundary Y. It is also required that D takes the form $\sigma(\partial/\partial u + A)$ in a neighbourhood of the boundary, where u is the normal parameter to the boundary, σ is a bundle isometry independent of u, and A is the obvious u-independent collar extension of an L^2 self-adjoint elliptic operator $A_0 : C^\infty(Y, E|_Y) \to C^\infty(Y, E|_Y)$. Note that A is automatically self-adjoint.

As we are dealing with sections over a manifold with boundary, it is appropriate to consider boundary conditions. It turns out that to obtain the index theorem we need to impose *global* boundary conditions. Consider the span of the eigenfunctions of A_0 which correspond to *negative* eigenvalues. This span is a subspace $S \subset C^\infty(Y, E|_Y)$. Our boundary condition is that we restrict the domain of the operator D to the subspace of $C^\infty(X, E)$ consisting of those sections f for which $f|_Y \in S$.

An alternative way of expressing this condition is to introduce the map

$$P : C^\infty(Y, E|_Y) \to C^\infty(Y, E|_Y)$$

which projects onto the span of the *non-negative* eigenfunctions. The boundary condition can then be written simply as $P(f|_Y) = 0$. With this in mind we will denote the new restricted domain of D (consisting of sections satisfying the boundary condition) by $C^\infty(X, E; P)$.

The idea now is to express the index of our first order elliptic differential operator D using traces of heat kernels, as explained in B.1 above, then use the asymptotic expansion as $t \to 0^+$ to compute the value of the index. So far this is the same strategy as can be employed for closed manifolds to prove the Atiyah-Singer index theorem. The difference with the current situation comes when trying to construct the heat kernel (fundamental solution) on a manifold with boundary. This is done by combining a fundamental solution on the closed manifold $X \cup_Y (-X)$, with a heat kernel on a cylinder (i.e. the middle region of the double) using suitable cut-off functions. Thus the overall heat kernel consists of two pieces: the 'interior of X piece' which is a cut-off from the heat kernel on the double of X, and the 'boundary piece' coming from the cylinder.

In detail, we start with the heat equation $\partial f/\partial u + D^*Df = 0$ on the manifold X, obtain a fundamental solution on the double of X produced in the usual way, and then restrict to one half. The asymptotic expansion for this heat kernel involves locally defined functions, and therefore such a restriction makes sense. Moreover the self-adjointness of A plays an important role in this: it means that D^*D and DD^* are isomorphic via σ in a neighbourhood of the boundary. Consequently the 'double X' contribution to the difference of the trace of the heat kernels vanishes near the boundary. When combined with a heat kernel from the cylinder, we obtain a decomposition as follows:

$$\mathrm{ind} D = \mathrm{tr} e^{-tD^*D} - \mathrm{tr} e^{-tDD^*} \sim K(t) + \sum_{k \geq -n} t^{k/2} \int_X \alpha_k(x)\, dx$$

where $K(t)$ is the cylinder (boundary) term, and the second term comes from the asymptotic expansion of the 'double X' contribution.

Atiyah, Patodi and Singer perform some detailed analysis on the difference of the heat kernels of D^*D and DD^* on the cylinder. Denoting this difference of kernels also by $K(t)$ they establish the following facts: a) setting $h = \dim\ker A$ we have $K(t) + h/2 \to 0$ exponentially as $t \to \infty$; b) $|K(t)| < Ct^{-n/2}$ as $t \to 0$ (i.e. $K(t)$ blows up, but it a controlled way); c) $K'(t) = (4\pi t)^{-1/2} \sum_k \lambda_k e^{-\lambda_k^2 t}$. Taken together, (a) and (b) show that the integral $\int_0^\infty (K(t)+h/2)t^{s-1}\,dt$, where s is a complex parameter, converges for all s with suitably large real part, and (c) then allows integration by parts. The importance of this integral is that the expression produced by performing the integration by parts involves the eta function. (It does not involve the parameter t: this gets integrated away.) Replacing $K(t)$ in the integral with its asymptotic expansion as $t \to 0^+$ yields a relationship between the coefficients in the expansion - which are of vital importance to the index formula as discussed in B.1 - and the eta function. Evaluating this expression at $s = 0$ yields the eta invariant on one side and a simple expression involving the index and $\int_X \alpha_0(x)\,dx$ on the other. This rearranges to give the Atiyah-Patodi-Singer index formula.

The function α_0 is the constant term in the asymptotic expansion of $\sum e^{-t\mu}|\phi_\mu(x)|^2 - \sum e^{-t\mu'}|\phi'_{\mu'}(x)|^2$ where the μ and ϕ_μ are the eigenvalues respectively eigenfunctions of D^*D, and μ' and $\phi'_{\mu'}$ are the eigenvalues respectively eigenfunctions of DD^* on the double of X. Note that by [Gi; Lemma 1.5] α_0 vanishes identically on odd dimensional closed manifolds. Applying this to the double of X shows that the $\int_X \alpha_0(x)\,dx$ term in the Atiyah-Patodi-Singer index formula vanishes in odd dimensions.

B.3 *Atiyah-Patodi-Singer: the collar approach*

Atiyah, Patodi and Singer also present an alternative approach to their index theorem which does not involve traces of heat kernels: rather, they obtain expressions for $\ker D$ and $\ker D^*$ directly by extending the manifolds with boundary to non-compact manifolds by adding collars. Specifically, they add an infinite collar $Y \times (-\infty, 0]$ to X, and consider the resulting non-compact manifold $\hat{X}$. The bundles and operators on X are then extended in the obvious way to $\hat{X}$. By expanding sections in the collar using the eigenfunctions of A_0 (corresponding to negative eigenvalues because of the boundary conditions), it follows easily that each section $f \in C^\infty(X, E; P)$ satisfying $Df = 0$ can be extended to a section over $\hat{X}$ by defining it as follows in the collar:

$$f(y, u) = \sum_{\lambda < 0} a_\lambda e^{-\lambda u} \phi_\lambda(y),$$

where ϕ_λ are the eigenfunctions of A_0 on Y, $u \in (-\infty, 0]$, and the a_λ are constants depending on f. This section decays quickly because of the exponential term, in which both λ and u are negative, and so this extended section is in L^2. The converse holds because the eigenfunctions not appearing in this expansion are not in L^2 (since they do not decay at all).

The situation for the adjoint D^* is slightly different. First we must consider what this adjoint actually is. It is not clear that there exists an operator D^* satisfying $(Df, \bar{f})_{L^2} = (f, D^*\bar{f})_{L^2}$ for all sections f, $\bar{f}$ of the bundles E respectively F over the non-compact manifold $\hat{X}$. In order to rectify this problem, we restrict to sections which are constant in the collar. As this means $\partial/\partial u\langle f, \bar{f}\rangle = 0$, we see that the pointwise adjoint of $\partial/\partial u$ in the collar is $-\partial/\partial u$. Given that A is self-adjoint, the adjoint of D in the collar is then easily seen to be $\sigma^{-1}(-\partial/\partial u + A)$.

The next issue we need to consider is the adjoint boundary condition. Recall that the original boundary condition on sections $f \in C^{\infty}(X, E)$ can be expressed as $Pf = 0$, where P denotes projection onto the non-negative eigenspaces. The adjoint boundary condition is then $(1-P)\bar{f} = 0$, i.e. the projection onto the *negative* eigenspaces is zero, so we restrict the domain of D^* to the subspace of sections of F belonging to the span of the non-negative eigenfunctions. The rationale behind this is as follows. We have $Pf = 0$ if and only if $(\sigma f, \bar{f}) = 0$ for all $\bar{f}$ in the span of the non-negative eigenfunctions of A. Therefore the original boundary condition is $(\sigma f, \bar{f}) = 0$ for all $\bar{f}$ such that $(1 - P)\bar{f} = 0$.

The essential difference between the original and the adjoint cases is that for $\bar{f} \in \ker D^*$, the expansion in the collar is $\bar{f}(y, u) = \sum_{\lambda \geq 0} a_\lambda e^{\lambda u}\phi_\lambda(y)$ where now $\lambda \geq 0$ because of the adjoint boundary conditions, but crucially this includes the possibility of $\lambda = 0$. We only get a $\lambda = 0$ eigenvalue if $\ker A \neq 0$, but if we do, the corresponding eigenfunctions do not decay. This leads to the concept of an *extended* L^2 section of F: a collar section which takes the form $\bar{f}(y, u) = \bar{g}(y, u) + \bar{f}_\infty(y)$ for large negative u, where $\bar{f}_\infty \in \ker A$ and $\bar{g}$ is in L^2.

The result is that $\operatorname{ind} D = h(E) - h(F) - h_\infty(F)$ where $h(E)$ and $h(F)$ are the dimensions of the space of of L^2-solutions to $Df = 0$ respectively $D^*\bar{f} = 0$, and $h_\infty(F)$ is the dimension of the subspace of $\ker A$ consisting of the limiting values of extended sections of F satisfying $D^*\bar{f} = 0$. Defining $h_\infty(E)$ in the analogous fashion to $h_\infty(F)$, in the special case of the Atiyah-Singer Dirac operator it can be shown that $h_\infty(E) + h_\infty(F)$ is equal to the dimension of the space of harmonic spinors. For the signature operator the various 'h' quantities can be interpreted via deRham cohomology to deduce the signature theorem for manifolds with boundary.

REFERENCES

[A] M. F. Atiyah, *K-theory*, (notes by D. W. Anderson), W. A. Benjamin inc., (1967).

[AB] K. Akutagawa, B. Botvinnik, *The relative Yamabe invariant*, Comm. Anal. Geom. **10** (2002), 925-954.

[ABK] P. Antonelli, D. Burghelea, P. J. Kahn, *Gromoll groups, Diff(S^n) and bilinear constructions of exotic spheres*, Bull. Amer. Math. Soc. **76** (1970), 772-777.

[ABS] M. F. Atiyah, R. Bott, A. Shapiro, *Clifford modules*, Topology **3** (1964), no. 1, 3-38.

[AM] U. Abresch, W. T. Meyer, *Pinching below 1/4, injectivity radius estimates, and sphere theorems*, J. Diff. Geom. **40** (1994), 643-691.

[APS1] M. F. Atiyah, V.K. Patodi, I. M. Singer, *Spectral asymmetry and Riemannian geometry I*, Math. Proc. Camb. Phil. Soc **77** (1975), 43-69.

[APS2] M. F. Atiyah, V.K. Patodi, I. M. Singer, *Spectral asymmetry and Riemannian geometry II*, Math. Proc. Camb. Phil. Soc **78** (1975), 405-432.

[AS3] M. F. Atiyah, I. M. Singer, *The index of elliptic operators III*, Ann. of Math. **87** (1968), 546-604.

[AS4] M. F. Atiyah, I. M. Singer, *The index of elliptic operators IV*, Ann. of Math. **93** (1971), 119-138.

[AS5] M. F. Atiyah, I. M. Singer, *The index of elliptic operators V*, Ann. of Math. **93** (1971), 139-148.

[AW] S. Aloff, N. Wallach, *An infinite family of distinct 7-manifolds admitting positively curved Riemannian structures*, Bull. Amer. Math. Soc. **81** (1975), 93-97.

[B1] B. Botvinnik, *Concordance and isotopy of metrics with positive scalar curvature*, Geom. Funct. Anal. **23** (2013), no. 4, 1099-1144.

[B2] B. Botvinnik, *Erratum to: Concordance and isotopy of metrics with positive scalar curvature*, Geom. Funct. Anal. **24** (2014), no. 3, 1037.

[Ba] Ya. V. Bazaikin, *On a certain family of closed 13-dimensional manifolds of positive curvature*, Siberian Mathematical Journal **37**, no. 6, (1996).

[Be] A.L. Besse, *Einstein Manifolds*, Springer-Verlag, Berlin (2002).

[BeEb] M. Berger, D. Ebin, *Some decompositions of the space of symmetric tensors on a Riemannian manifold*, J. Diff. Geom. **3** (1969), 379-392.

[BeGu] V. Berestovski, L. Guijarro, *A metric characterization of Riemannian submersions*, Ann. Global Anal. Geom. **18** (2000), 577-588.

[Bel] I. Belegradek, *Vector bundles with infinitely many souls*, Proc. Amer. Math. Soc. **131** (2003), 2217-2221.

W. Tuschmann, D.J. Wraith, *Moduli Spaces of Riemannian Metrics*,
Oberwolfach Seminars 46, DOI 10.1007/978-3-0348-0948-1

[BER] B. Botvinnik, J. Ebert, O. Randal-Williams, *Infinite loop spaces and positive scalar curvature*, arxiv:1411.7408.

[BF] C. Blanc, F. Fiala, *Le type d'une surface et sa courbure totale*, Comment. Math. Helv. **14**, (1941-42), 230-233.

[BFK] I. Belegradek, F.T. Farrell, V. Kapovitch, *Space of non-negatively curved manifolds and pseudoisotopies*, arXiv:1501.03475.

[BG] B. Botvinnik, P. Gilkey, *The eta invariant and metrics of positive scalar curvature*, Math. Ann. 302 (1995), no. 3, 507-517.

[BH] I. Belegradek and J. Hu, *Connectedness properties of the space of complete non-negatively curved planes*, Math. Ann. **362** (2015), no. 3-4, 1273-1286.

[BHSW] B. Botvinnik, B. Hanke, T. Schick, M. Walsh, *Homotopy groups of the moduli space of metrics of positive scalar curvature*, Geom. Topol. **14** (2010), 2047-2076.

[BKS] I. Belegradek, S. Kwasik, R. Schultz, *Moduli spaces of non-negative sectional curvature and non-unique souls*, J. Diff. Geom. **89** (2011), 49-86.

[Bl] D. E. Blair, *Spaces of metrics and curvature functionals*, Handbook of Differential Geometry I, North Holland, Amsterdam (2000), 153-185.

[BoKl] A. I. Bobenko, Ch. Klein (Eds.), *Computational Approach to Riemann Surfaces*, Lecture Notes in Mathematics **2013**, Springer-Verlag (2011).

[Br] W. Browder, *Surgery on simply connected manifolds*, Springer-Verlag, Berlin, (1972).

[BT] Y. Burago, V. A. Toponogov, *On three-dimensional Riemannian spaces with curvature bounded above*, Matematicheskie Zametki **13** (1973), 881-887.

[BV] J. M. Boardman, R. M. Vogt, *Homotopy invariant algebraic structures on topological spaces*, Lecture Notes in Mathematics, **347**. Springer-Verlag, (1973).

[Ca] R. Carr, *Construction of manifolds of positive scalar curvature*, Trans. Amer. Math. Soc. **307** (1988), no. 1, 63-74.

[Ce1] J. Cerf, *Sur les difféomorphismes de la sphère de dimension trois* ($\Gamma_4 = 0$), Lecture Notes in Math. **53**, Springer-Verlag, (1968).

[Ce2] J. Cerf, *La stratification naturelle des espaces de fonctions différentiables réelles et la théorème de la pseudo-isotopie*, Inst. Hautes Etudes Sci. Publ. Math. **39** (1970), 5-173.

[CG1] J. Cheeger, D. Gromoll, *On the structure of complete manifolds of non-negative curvature*, Ann. of Math. **96** (1972), 413-443.

[CG2] J. Cheeger, D. Gromoll, *On the lower bound for the injectivity radius of* $1/4$*-pinched Riemannian manifolds*, J. Diff. Geom. **15** (1980), 437-442.

[Ch] V. Chernysh, *On the homotopy type of the space* $\mathcal{R}^+(M)$, arxiv:math.GT/0405235.

[Cl] B. Clarke, *The completion of the manifold of Riemannian metrics with respect to its L^2 metric*, arXiv:0904.0159v1.

[CLN] B. Chow, P. Lu, L. Ni, *Hamilton's Ricci flow*, Graduate Studies in Mathematics **77**, American Mathematical Society, Providence, RI; Science Press, New York (2006).

[CM] F. Codá Marques, *Deforming three-manifolds with positive scalar curvature*, Ann. of Math. (2) **176** (2012), no. 2, 815-863.

[CS] D. Crowley, T. Schick, *The Gromoll filtration, KO-characteristic classes and metrics of positive scalar curvature*, Geom. Topol. **17** (2013), 1773-1790.

[CW] D. Crowley, D. J. Wraith, *Positive Ricci curvature on highly connected manifolds*, arXiv:1404.7446.v1, J. Diff. Geom. (to appear).

[deR] G. de Rham, *Complexes à automorphismes et homéomorphie différentiable*, Ann. Inst. Fourier Grenoble **2**, (1950), 51-67.

[doC] M. do Carmo, *Riemannian Geometry*, (third edition), Birkhäuser, (1992).

[Eb] D. G. Ebin, *The manifold of Riemannian metrics*, Global Analysis (Proc. Sympos. Pure Math., Vol. XV, Berkeley, Calif., 1968), Amer. Math. Soc., Providence, R.I., (1970), 11-40.

[EK] J. Eells, N. Kuiper, *An invariant for certain smooth manifolds*, Annali di Math. **60** (1962), 93-110.

[Es] J.-H Eschenburg, *New examples of manifolds with strictly positive curvature*, Invent. math. **66** (1982), 469-480.

[FH] F. T. Farrell, W. C. Hsiang, *On the rational homotopy groups of the diffeomorphism groups of discs, spheres and aspherical manifolds*, Proc. Sympos. Pure Math. XXXII, Amer. Math. Soc., Providence, R.I., (1978).

[FG] D. S. Freed, D. Groisser, *The basic geometry of the manifold of Riemannian metrics and of its quotient by the diffeomorphism group*, Michigan Math. J. **36** (1989), 323-344.

[FM] H. D. Fegan, R. S. Millman, *Quadrants of Riemannian metrics*, Michigan Math. J. **25** (1978), no. 1, 3-7.

[FO1] F. T. Farrell, P. Ontaneda, *The Teichmüller space of pinched negatively curved metrics on a hyperbolic manifold is not contractible*, Ann. of Math. (2) **170** (2009), no. 1, 45-65.

[FO2] F. T. Farrell, P. Ontaneda, *The moduli space of negatively curved metrics of a hyperbolic manifold*, J. Topol. **3** (2010), no. 3, 561-577.

[FO3] F. T. Farrell, P. Ontaneda, *On the topology of the space of negatively curved metrics*, J. Diff. Geom. **86** (2010), no. 2, 273-301.

[FR] F. Fang, X. Rong, *Positive Pinching, Volume and Second Betti Number*, Geometric and Functional Analysis (GAFA) **9** (1999), 641-674.

[Ga] P. Gajer, *Riemannian metrics of positive scalar curvature on compact manifolds with boundary*, Ann. Global Anal. Geom. **5** (1987), no. 3, 179-191.

[GG] M. Golubitsky, V. Guillemin, *Stable mappings and their singularities*, Graduate Texts in Mathematics **14**, Springer-Verlag, (1973).

[Gi] P. Gilkey, *The η-invariant for even dimensional pin^c-manifolds*, Adv. in Math. **58** (1985), 243-284.

[GL1] M. Gromov and H.B. Lawson, *The classification of manifolds of positive scalar curvature*, Ann. of Math. **111** (1980), 423-434.

[GL2] M. Gromov, H.B. Lawson, *Positive scalar curvature and the Dirac operator on complete Riemannian manifolds*, Publ. Math. Inst. Hautes Etudes Sci. **58** (1983), 83-196.

[GM] O. Gil-Medrano, P. W. Michor, *The Riemannian manifold of all Riemannian metrics*, Quart. J. Math. Oxford Ser. (2) **42** (1991), no. 166, 183-202.

[Go] S. Goette, *Morse Theory and higher torsion invariants I*, math.DG/0111222.

[Gui] L. Guijarro, *On the metric structure of open manifolds with non-negative curvature*, Pacific J. Math. **196** (2000), 429-444.

[GZ] K. Grove, W. Ziller, *Curvature and Symmetry of Milnor Spheres*, Ann. of Math. **152** (2000), 331-367.

[H] M. Hirsch, *Differential Topology*, Graduate Texts in Mathematics, Springer-Verlag, (1997).

[Ha] A. E. Hatcher, *A proof of the Smale conjecture, Diff(S^3) $\simeq$ $O(4)$*, Ann. of Math. (2) **117** (1983), no. 3, 553-607.

[Hae] A. Haefliger, *Plongements différentiables de variétés dans variétés*, Comment. Math. Helv. **36** (1961), 47-82.

[Ham] R. Hamilton, *The inverse function theorem of Nash and Moser*, Bull. Am. Math. Soc **7** (1982), 65-222.

[Hi] N. Hitchin, *Harmonic Spinors*, Adv. in Math. **14** (1974), 1-55.

[HSS] B. Hanke, T. Schick, W. Steimle, *The space of metrics of positive scalar curvature*, Publ. Math. Inst. Hautes Études Sci. **120** (2014), 335-367.

[Hu] D. Husemoller, *Fibre Bundles*, Springer-Verlag, third edition (1994).

[Kl1] W. Klingenberg, *Contributions to Riemannian geometry in the large*, Ann. of Math. **69**, (1959), 654-666.

[Kl2] W. Klingenberg, *Über Riemannsche Mannigfaltigkeiten mit positiver Krümmung*, Comm. Math. Helv. **35** (1961), 47-54.

[KlSa1] W. Klingenberg, T. Sakai, *Injectivity radius estimates for 1/4-pinched manifolds*, Arch. Math. **34** (1980), 371-376.

[KlSa2] W. Klingenberg, T. Sakai, *Remarks on the injectivity radius estimate for almost* 1/4-*pinched manifolds*, Lecture Notes in Math. **1201**, Springer-Verlag (1986).

[KM] M. Kervaire, J. Milnor, *Groups of homotopy spheres*, Ann. of Math. **77** (1963), 504-537.

[KPT] V. Kapovitch, A. Petrunin, W. Tuschmann, *Non-negative pinching, moduli spaces and bundles with infinitely many souls*, J. Diff. Geom. **71** (2005) no. 3, 365-383.

[KrMi] A. Kriegl, P. Michor, *The convenient setting of global analysis*, Mathematical Surveys and Monographs **53**, American Mathematical Society (1997).

[KS1] M. Kreck, S. Stolz, *A diffeomorphism classification of 7-dimensional homogeneous Einstein manifolds with* $SU(3) \times SU(2) \times U(1)$*-symmetry*, Ann. of Math. (2) **127** (1988), 373-388.

[KS2] M. Kreck, S. Stolz, *Some nondiffeomorphic homeomorphic homogeneous 7-manifolds with positive sectional curvature*, J. Diff. Geom. **33** (1991), 465-486.

[KS3] M. Kreck, S. Stolz, *Nonconnected moduli spaces of positive sectional curvature metrics*, J. Am. Math. Soc. **6** (1993), 825-850.

[L1] J. Lohkamp, *Metrics of negative Ricci curvature*, Ann. of Math. **140** (1994), 655-683.

[L2] J. Lohkamp, *Negative bending of open manifolds*, J. Diff. Geom. **40** (1994), 461-474.

[L3] J. Lohkamp, *Curvature h-principles*, Ann. of Math. **142** (1995), 457-498.

[LM] H.B. Lawson, M.-L. Michelsohn, *Spin Geometry*, Princeton Math. Series **38**, Princeton University Press, (1989).

[LuMi] G. Luke, A. S. Mishchenko, *Vector Bundles and Their Applications*, Springer (1998).

[M1] J. Milnor, *A unique decomposition theorem for 3-manifolds*, Amer. J. Math. **84**, (1962), 1-7.

[M2] J. Milnor, *Remarks concerning spin manifolds*, Differential and Combinatorial Topology (A symposium in honor of Marston Morse), Princeton Univ. Press, (1965), 55-62.

[Ma] J. P. May, *The geometry of iterated loop spaces*, Lectures Notes in Mathematics, **271**, Springer-Verlag, (1972).

[Mi] T. Miyazaki, *On the existence of positive curvature metrics on non-simply-connected manifolds*, J. Fac. Sci. Univ. Tokyo, Sect. IA **30** (1984), 549-561.

[Mo] S. Morita, *Geometry of Differential Forms*, Translations of mathematical monographs **201**, Amer. Math. Soc. (2001).

[Mos] G. D. Mostow, *Quasi-conformal mappings in n-space and the rigidity of the hyperbolic space forms*, Publ. Math. IHES **34** (1968), 53-104.

[MS] J. Milnor, J. Stasheff, *Characteristic Classes*, Annals of Mathematical Studies **76**, Princeton Univ. Press (1974).

[Pa] R. S. Palais, *Homotopy theory of infinite dimensional manifolds*, Topology **5**, (1966), 1-16.

[Per] G. Perelman, *Proof of the soul conjecture of Cheeger and Gromoll*, J. Diff. Geom. **40** (1994), 209-212.

[PRT] A. Petrunin, X. Rong, W. Tuschmann, *Collapsing versus Positive Pinching*, Geometric and Functional Analysis (GAFA) **9** (1999), 699-735.

[Po] A. Pogorelov, *A theorem regarding geodesics on closed convex surfaces*, Rec. Math. (Mat. Sbornik) N.S. **18** (**60**) (1946), 181-183.

[PT] A. Petrunin, W. Tuschmann, *Diffeomorphism Finiteness, Positive Pinching, and Second Homotopy*, Geometric and Functional Analysis (GAFA) **9** (1999), 736-774.

[Pü] T. Püttmann, *Optimal pinching constants of odd dimensional homogeneous spaces*, Invent. math. **138** (1999), 631-684.

[Ra] A. A. Ranicki, *Algebraic and Geometric Surgery*, Oxford University Press, (2002).

[Ro] J. Rosenberg, *C^*-algebras, positive scalar curvature, and the Novikov conjecture, II*, Geometric Methods in Operator Algebras, Pitman Res. Notes **123**, Longman, (1986) 341-374.

[Sa] T. Sakai, *On a theorem of Burago-Toponogov*, Indiana Univ. Math. J. **32** (1983), 165-175.

[Sm] S. Smale, *On the structure of* 5-*manifolds*, Ann. of Math. **74** (1962), 38-46.

[St] S. Stolz, *Simply connected manifolds with positive scalar curvature*, Ann. of Math. **136** (1992), 511-540.

[SY] J. P. Sha and D. G. Yang, *Positive Ricci curvature on the connected sums of $S^n \times S^m$*, J. Diff. Geom. **33** (1990), 127-138.

[Ro] J. Rosenberg, *Manifolds of positive scalar curvature: a progress report*, Surveys in Differential Geometry XI (J. Cheeger and K. Grove eds.), International Press (2007), 259-294.

[RS] J. Rosenberg and S. Stolz, *Metrics of positive scalar curvature and connections with surgery*, Surveys on Surgery Theory vol. 2, Annals of Mathematical Studies **149**, Princeton University Press, (2001), 353-386.

[Ru] W. Rudin, *Functional Analysis*, McGraw-Hill (1973).

[Se] R. Seeley, *Complex powers of an elliptic operator*, Proc. Symp. Pure Math., Amer. Math. Soc. **10** (1967), 288-307.

[Sha1] V. A. Sharafutdinov, *The Pogorelov-Klingenberg theorem for manifolds homeomorphic to R^n*, Siberian Mathematical Journal **18** (1977), 649-657.

[Sha2] V. A. Sharafutdinov, *Convex sets in a manifold of nonnegative curvature*, Mat. Zametki **26** no. 1 (1979), 129-136. (1979), 56-560.

[Sie] L. C. Siebenmann, *On detecting open collars*, Trans. Am. Math. Soc. **142** (1969), 201-227.

[V] R. M. Vogt, *Cofibrant operads and universal E_∞ operads*, Topology Appl. **133** (2003), no. 1, 69-87.

[W1] C. T. C. Wall, *Classification of $(n-1)$-connected $2n$-manifolds.*, Ann. of Math., **75** (1962), 163-189.

[W2] C. T. C. Wall, *Classification problems in differential topology - VI*, Topology, **6** (1967), 273-296.

[Wa1] M. Walsh, *Metrics of positive scalar curvature and generalized Morse functions, Part I*, Mem. Amer. Math. Soc. **209** (2011), no. 983.

[Wa2] M. Walsh, *Cobordism invariance of the homotopy type of the space of positive scalar curvature metrics*, Proc. Amer. Math. Soc. **141** (2013), no. 7, 2475-2484.

[Wa3] M. Walsh, *H-spaces, loop spaces and the space of positive scalar curvature metrics on the sphere*, Geom. Topol. **18** (2014), no. 4, 2189-2243.

[Wi] D. Wilkens, *Closed $(s-1)$-connected $(2s+1)$-manifolds, $s = 3, 7$*, Bull. London Math. Soc. **4** (1972), 27-31.

[Wil] B. Wilking, *A duality theorem for Riemannian foliations in nonnegative sectional curvature*, Geom. Funct. Anal. **17** (2007), 1297-1320.

[Wr1] D. J. Wraith, *Exotic spheres with positive Ricci curvature*, J. Diff. Geom. **45** (1997), 638-649.

[Wr2] D. J. Wraith, *Surgery on Ricci positive manifolds*, J. reine angew. Math. **501** (1998), 99-113.

[Wr3] D. J. Wraith, *New connected sums with positive Ricci curvature*, Ann. Glob. Anal. Geom. **32** (2007), 343-360.

[Wr4] D. J. Wraith, *On the moduli space of positive Ricci curvature metrics on homotopy spheres*, Geom. Topol. **15** (2011), 1983-2015.

[WZ] M. Wang, W. Ziller, *Einstein metrics with positive scalar curvature*, Curvature and Topology of Riemannian Manifolds, Proceedings, Lecture Notes in Math. **1021**, Springer-Verlag (1986), 319-336.

[Yam] T. Yamaguchi, *Collapsing and pinching under a lower curvature bound*, Ann. of Math. (2) **133** (1991), 317-357.

[Yau] S.-T. Yau, *Problem Section. Seminar on differential geometry*, Ann. Math. Stud. **102** (1982).

Printed in the United States
By Bookmasters